The Cambridge Pre-GED Program in MATH

1988 Edition

CAMBRIDGE Adult Education
Prentice Hall Regents, Englewood Cliffs, NJ 07632

**The Cambridge Pre-GED Program in
MATH**

Library of Congress Cataloging-in-Publication Data

The Cambridge pre-GED program in math.

 1. Mathematics--Examinations, questions, etc.
 2. General educational development tests.
 QA43.C198 1988   510'.76   88-9564
 ISBN 0-13-113762-X

Editorial supervision: Tim Foote
Production supervision: Janet Johnston
Manufacturing buyer: Art Michalez

Some of the material in this book originally appeared in *Introduction to Arithmetic*
Angelica W. Cass, general editor, copyright 1972 by Cambridge Book Company.

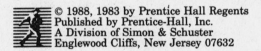
Printed in the United States of America

10 9 8 7 6 5 4 3

ISBN 0-13-113762-X

Prentice-Hall International (UK) Limited, *London*
Prentice-Hall of Australia Pty. Limited, *Sydney*
Prentice-Hall Canada Inc., *Toronto*
Prentice-Hall Hispanoamericana, S.A., *Mexico*
Prentice-Hall of India Private Limited, *New Delhi*
Prentice-Hall of Japan, Inc., *Tokyo*
Simon & Schuster Asia Pte. Ltd., *Singapore*
Editora Prentice-Hall do Brasil, Ltda., *Rio de Janeiro*

CONTENTS

v

INTRODUCTION

The forty chapters in this book will help you begin your preparation to take the GED, or the High School Equivalency Exam, as it is sometimes called. The Mathematics Test of the GED is made up of 56 problems. Half of the problems are in arithmetic; half are algebra and geometry problems. Almost all the arithmetic problems are word problems, and some of them are based on charts and graphs. You can use the instruction and practice in this book to help you develop *all* the computation skills you need for success with the test.

This book has several features—you will find most of them listed in the Contents on pages iii to v—to help you get the most out of your study:

- The **Pretest** will help you find out about your current skills in arithmetic.
- Each of the five parts of this book begins with the exercise **How Much Do You Know Already?** to help you decide which chapters you should study.
- The five exercises called **Measuring Your Progress** and the **Posttest** help you find out how your skills have improved and which chapters, if any, you should review.
- Throughout this book there are sets of practice problems called **Try It**, hints for solving word problems, and answer sections to help you check your work. Among the practice problems are nearly 500 word and graphic items to help you build the problem-solving skills the Mathematics Test requires.

You can use the *Cambridge Pre-GED Exercise Book in Math* for extra practice as you work through this textbook. The charts on the inside covers of this book and the exercise book show which pages in the exercise book are related to the chapters in this book.

When you have successfully completed all the chapters in this book, you will have developed all the computation skills and some of the problem-solving skills you will need as you continue your preparation for the GED. You will be able to move right into the math instruction in Cambridge's GED books because the chapters on whole numbers, fractions, decimals, percents, and graphs build from reviews of the skills you will have learned using this book.

PRETEST

This test is made up of 40 problems similar to the ones in this book. Solve all the problems you are able to. Then check your answers against those on page xi. The results will help you find out where you should begin your study in this book.

1. What is the value of the 4 in the number 5,407?

2. 120
   23
   +46

3. 3,643
   +2,778

4. 4,764
   −520

5. 5,204
   −3,396

6. 163
   ×3

7. 5,404
   × 54

8. $4\overline{)804}$

9. $24\overline{)1920}$

10. Write the mixed number that equals $\frac{17}{3}$.

11. $2\frac{2}{5}$
    $+ \frac{3}{5}$
    _____

12. $3 - 1\frac{7}{8} =$

13. $\frac{2}{3} + \frac{1}{4} =$

14. $1\frac{1}{2} \times \frac{1}{6} =$

15. $\frac{1}{2} \div 2\frac{1}{6} =$

16. Write $20\frac{86}{100}$ as a decimal.

17. $.44 + .016 + 1.1 =$

18. $10.25
    − 8.75

19. $.7 \times 3.14 =$

20. $1.01\overline{)80.8}$

21. Write $\frac{1}{4}$ as a percent.

22. 8% of 50 = ___.

23. 75 = ___% of 30.

24. 17 = $33\frac{1}{3}$% of ___.

25. Find the interest on a 1-year loan of $1200 at 18%.

	Bel Air	Birchwood City	Mason City	Chisholm	Salina
Bel Air	–	215	174	388	109
Birchwood City	215	–	74	189	106
Mason City	174	74	–	263	59
Chisholm	388	189	263	–	165
Salina	109	106	59	165	–

26. Using the chart (right), how many miles is it from Mason City to Chisholm?

27. According to the circle graph (right), how much does this family spend on food, rent, and gasoline combined?

Family's Monthly Budget

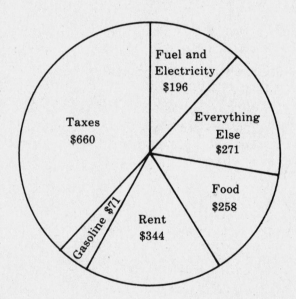

28. Based on the pictograph, how much greater is Ohio's enrollment increase than Illinois'?

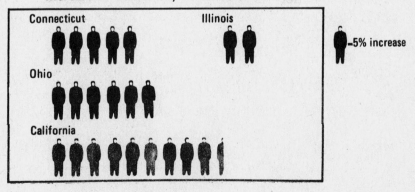

Increase in Elementary School Enrollments

29. According to the bar graph below, during which two years was Evanstown's population the same?

POPULATION OF EVANSTOWN, 1900–1980

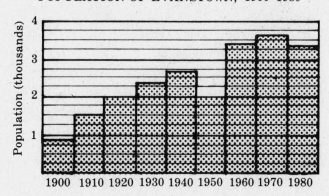

30. In what year did the trend in the line graph (right) change directions?

AVERAGE NUMBER OF EMPLOYEES IN THE MINING INDUSTRY, 1955–1976

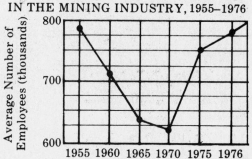

31. Lee was laid off from his job in the fall this year. He made $6,187 less than he did last year. This year he made $13,574. What was his income last year?

32. Minerva wants to move with her children to a new apartment. One landlord told her the annual lease for a 2-bedroom apartment would be $4260. How much rent would she have to pay per month?

33. Willie bought a two-by-four that was 10 feet long. He used only $8\frac{7}{8}$ feet. How long was the piece left over?

34. Hector has $2\frac{1}{2}$ cups of rice. He figures that it takes about $\frac{1}{2}$ cup of rice to feed one person. How many people can he feed with the rice he has?

35. Loretta noticed that her car's odometer showed 10,964.8 miles before she drove to her brother's house. When she got home again, it showed 11,126.3 miles. How far had she driven?

36. A car rental agency charges $.16 for each mile driven. If James drives 249.5 miles, how much must he pay for mileage?

37. At the adult learning center, 80% of the students in the evening class completed the GED course. In the day class, $\frac{3}{4}$ of the students finished the course. Which class had the higher course completion rate?

38. Only 12 out of 20 people who start a certain diet lose the weight they want to. What percent of the people who start the diet reach their goal?

39. According to the circle graph below, which two of the family's expenses together are exactly 50% of the budget?

40. Based on the circle graph, if the family's weekly income is $350, how much is the weekly telephone expense?

Family Budget

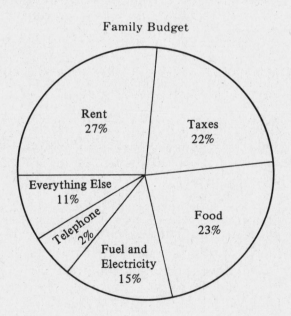

PRETEST ANSWERS

Put an X by each problem you did wrong or did not do. If you missed one or more problems in any group, do the *How Much Do You Know Already?* exercise on the page indicated.

Group	Computation Problems	Word Problems	Exercise on Page
Whole Numbers	1. **400** 2. **189** 3. **6,421** 4. **4,244** 5. **1,808** 6. **489** 7. **291,816** 8. **201** 9. **80**	31. **$19,761** 32. **$355**	1
Fractions	10. $5\frac{2}{3}$ 11. **3** 12. $1\frac{1}{8}$ 13. $\frac{11}{12}$ 14. $\frac{1}{4}$ 15. $\frac{3}{13}$	33. $1\frac{1}{8}$ **feet** 34. **5 people**	94
Decimals	16. **20.86** 17. **1.556** 18. **$1.50** 19. **2.198** 20. **80**	35. **161.5 miles** 36. **$39.92**	147
Percents	21. **25%** 22. **4** 23. **250%** 24. **51** 25. **$216**	37. **Evening** 38. **60%**	190
Charts and Graphs	26. **263** 27. **$673** 28. **20%** 29. **1920 and 1950** 30. **1970**	39. **Rent and food** 40. **$7.00**	239

You will find instruction about problems like those on the Pretest in the chapters indicated in the following table.

Problem	Chapter	Problem	Chapter	Problem	Chapter
1	2	11	14	21, 37	28
2	3	12, 33	15	22	29
3, 31	4	13	16	23, 38	30
4	5	14	17	24	31
5	6	15, 34	18	25	32
6	7	16	21	26	35
7	8	17	22	27, 39, 40	36
8	9	18, 35	23	28	37
9, 32	10	19, 36	24	29	38
10	13	20	25	30	39

SOLVING WORD PROBLEMS

There are three basic differences between the way computation problems and word problems are presented. Compare these two problems:

Computation Problem	Word Problem
$ 2.25 + .75 [ ]	Manuel went shopping and bought paper cups for $2.25 and plastic spoons for $.75. How much did he pay altogether for these items?

Both problems require the same calculation. The difference between them is in the way the information is presented to you. In the computation problem, you can see three things immediately:

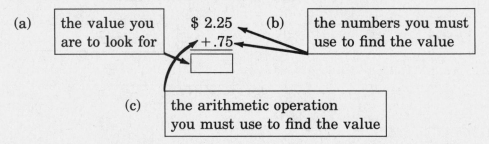

(a) the value you are to look for — $ 2.25   (b) the numbers you must use to find the value
+ .75
[   ]

(c) the arithmetic operation you must use to find the value

In the word problem, you must *find out* those same three items: (a) the value you are to look for, (b) the numbers to use, and (c) the operation to use. Part of problem solving is finding out *what* it is you want to know and then figuring out *how* to find that.

Follow these six steps when you are solving word problems:

Steps for Solving Word Problems

1. Read the problem *carefully*. Pay attention to every word.
2. Figure out what value you should find. Often the last sentence of a problem is the best clue.
3. Decide which numbers in the problem to use to find the value you need.
4. Decide whether to add, subtract, multiply, or divide. There may be clue words in the problem to help you decide.
5. Solve the problem.
6. Check your arithmetic and ask yourself whether your answer makes sense.

You will find examples of clue words for the four arithmetic operations on the following pages:

addition—page 16 or 155   multiplication—page 46 or 165
subtraction—page 29 or 160   division—page 63 or 175

Whole Numbers

1. HOW MUCH DO YOU KNOW ALREADY?

Here are some problems with whole numbers. You will be asked to add, subtract, multiply, or divide. This will help you find out how much you know already. You will also find out what you need to study. Do as many problems as you can. Write your answers on the lines at the right. Then, turn to page 77. Check your answers. Page 77 will tell you what pages in the book to study.

1.
$$\begin{array}{r} 4 \\ + 5 \\ \hline \end{array}$$

2. 6+3=

3.
$$\begin{array}{r} 34 \\ + 55 \\ \hline \end{array}$$

4.
$$\begin{array}{r} 245 \\ + 112 \\ \hline \end{array}$$

5.
$$\begin{array}{r} 546 \\ + 31 \\ \hline \end{array}$$

6.
$$\begin{array}{r} 1 \\ 3 \\ + 2 \\ \hline \end{array}$$

7.
$$\begin{array}{r} 2 \\ 4 \\ 0 \\ 1 \\ + 2 \\ \hline \end{array}$$

8.
$$\begin{array}{r} 13 \\ 23 \\ + 51 \\ \hline \end{array}$$

9.
$$\begin{array}{r} 634 \\ 21 \\ 310 \\ + 4 \\ \hline \end{array}$$

10.
$$\begin{array}{r} 7 \\ + 8 \\ \hline \end{array}$$

11.
$$\begin{array}{r} 63 \\ + 72 \\ \hline \end{array}$$

12.
$$\begin{array}{r} 38 \\ + 47 \\ \hline \end{array}$$

13.
$$\begin{array}{r} 66 \\ + 4 \\ \hline \end{array}$$

14.
$$\begin{array}{r} 483 \\ + 361 \\ \hline \end{array}$$

15.
$$\begin{array}{r} 4,732 \\ + 3,624 \\ \hline \end{array}$$

1. _____
2. _____
3. _____
4. _____
5. _____
6. _____
7. _____
8. _____
9. _____
10. _____
11. _____
12. _____
13. _____
14. _____
15. _____

1

2 WHOLE NUMBERS

16.   7,684
     + 1,496

17.   2,001
     + 1,999

18.   3,478
      5,963
      8,109
     + 7,575

19.   5
    − 2

20. 6 − 4 =

21.   26
    − 4

22.   263
    − 142

23.   4,602
    − 601

24.   8,648
    − 27

25.   8,278
    − 434

26.   309
    − 127

27.   5,734
    − 2,868

28.   703
    − 534

29.   8,006
    − 2,629

30.   10,273
    − 565

31.   8
    × 9

32.   36
    × 2

33.   9
    × 0

34.   40
    × 2

35.   146
    × 2

36.   2,675
    × 3

16. _____
17. _____
18. _____
19. _____
20. _____
21. _____
22. _____
23. _____
24. _____
25. _____
26. _____
27. _____
28. _____
29. _____
30. _____
31. _____
32. _____
33. _____
34. _____
35. _____
36. _____

37.   978
      × 68

38.   307
      × 29

39.   3,072
      ×   85

40.   8,093
      ×   527

41.   244
      × 102

42.   7,803
      ×   978

43. 8$\overline{)72}$

44. 3$\overline{)39}$

45. 3$\overline{)90}$

46. 3$\overline{)306}$

47. 3$\overline{)123}$

48. 4$\overline{)35}$

49. 12$\overline{)38}$

50. 43$\overline{)129}$

51. 32$\overline{)992}$

52. 34$\overline{)3,808}$

53. 325$\overline{)6,847}$

54. 23$\overline{)4,899}$

37. _____
38. _____
39. _____
40. _____
41. _____
42. _____
43. _____
44. _____
45. _____
46. _____
47. _____
48. _____
49. _____
50. _____
51. _____
52. _____
53. _____
54. _____

2. HOW MUCH IS A NUMBER WORTH?

Everyone uses numbers, everywhere, every day. How much money do you earn each week? How many people are in your family? How many people live in your city? How many miles do you drive to work? How much gas does your car use? The answers to these questions are numbers.

To make sense out of the answers, you must know the value of numbers. Would you rather earn $130 or $13? The 130 is worth more than 13. Why? What is the difference between these two numbers? To tell how much a number is worth, look at how many places are in the number. 130 has three places. 13 has only two places. 130 is worth more because it has more places.

PLACE VALUE

Both people and numbers have names. A person's name tells you who he is. The name of a number tells you how much it is worth. You read the names of numbers by reading the names of the places in the number. The digits 0, 1, 2, 3, 4, 5, 6, 7, 8, and 9 are all you need to write any number. The value of these digits depends on their place in a number.

The first place is named ONES. How many ONES are there in the number 2?

$$1 \quad 1 \quad = \quad 2 \text{ ones}$$

There are two ones in the number 2.

The 2 can have different values. Imagine you have a pile of pennies. How do you find out how many pennies are in the pile? You could put the pennies in stacks of tens. If you have two stacks, you have 20 pennies. The number 20 is made up of 2 tens and 0 ones. TENS is the next place after ones.

We count money in tens every day. A dime is worth 10 pennies. If you have 4 dimes and 3 pennies in your pocket, you have 43¢.

4

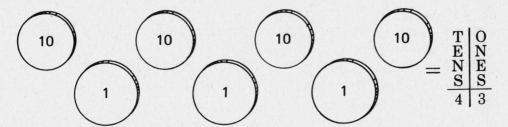

43¢ has 4 tens and 3 ones.

Try It (Answers on page 78.)

How many tens and ones are in each of these numbers?

1. 42 has_____tens and_____ones.

2. 89 has_____tens and_____ones.

3. 31 has_____tens and_____ones.

4. 55 has_____tens and_____ones.

5. 93 has_____tens and_____ones.

6. 67 has_____tens and_____ones.

7. 28 has_____tens and_____ones.

8. 16 has_____tens and_____ones.

9. 74 has_____tens and_____ones.

10. 92 has_____tens and_____ones.

Each place in our system is worth ten groups of the place to its right. Ten ones is the same as one ten. Ten groups of ten are one hundred. The third place in our system is HUNDREDS.

 "This pill contains hundreds of tiny time capsules." If the pill has 483 tiny time capsules, the capsules might be grouped like this:

HUNDREDS

TENS

ONES

The number 483 has 4 hundreds, 8 tens, and 3 ones.

Try It (Answers on page 78.)

1. 689 has_____hundreds,_____tens, and_____ones.

2. 471 has_____hundreds,_____tens, and_____ones.

3. 513 has_____hundreds,_____tens, and_____ones.

4. 837 has_____hundreds,_____tens, and_____ones.

5. 125 has_____hundreds,_____tens, and_____ones.

6. 952 has_____hundreds,_____tens, and_____ones.

7. 264 has_____hundreds,_____tens, and_____ones.

8. 398 has_____hundreds,_____tens, and_____ones.

9. 612 has_____hundreds,_____tens, and_____ones.

10. 746 has_____hundreds,_____tens, and_____ones.

Babe Ruth went to bat eight thousand three hundred ninety-nine times (8,399). The number 8,399 contains the fourth place in value: THOUSANDS.

8,399 looks like this:

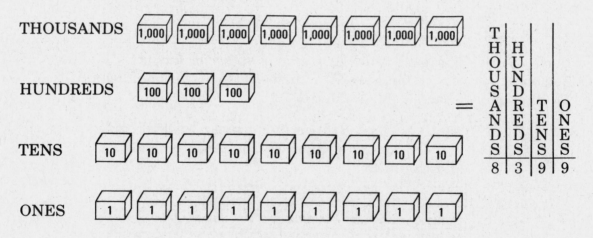

8,399 has 8 thousands, 3 hundreds, 9 tens, and 9 ones.

Try It (Answers on page 78.)

Fill in the number of thousands, hundreds, tens, and ones in the numbers below.

1. 1,535 has_____thousands,_____hundreds,_____tens, and_____ones.

2. 6,421 has＿＿＿thousands,＿＿＿hundreds,＿＿＿tens, and＿＿＿ones.

3. 3,789 has＿＿＿thousands,＿＿＿hundreds,＿＿＿tens, and＿＿＿ones.

4. 4,618 has＿＿＿thousands,＿＿＿hundreds,＿＿＿tens, and＿＿＿ones.

5. 7,296 has＿＿＿thousands,＿＿＿hundreds,＿＿＿tens, and＿＿＿ones.

6. 5,812 has＿＿＿thousands,＿＿＿hundreds,＿＿＿tens, and＿＿＿ones.

7. 9,374 has＿＿＿thousands,＿＿＿hundreds,＿＿＿tens, and＿＿＿ones.

8. 4,965 has＿＿＿thousands,＿＿＿hundreds,＿＿＿tens, and＿＿＿ones.

9. 8,147 has＿＿＿thousands,＿＿＿hundreds,＿＿＿tens, and＿＿＿ones.

10. 2,953 has＿＿＿thousands,＿＿＿hundreds,＿＿＿tens, and＿＿＿ones.

The next place after thousands is called TEN THOUSANDS. Renata smokes a pack of cigarettes a day. She smokes about 73,143 cigarettes in 10 years.

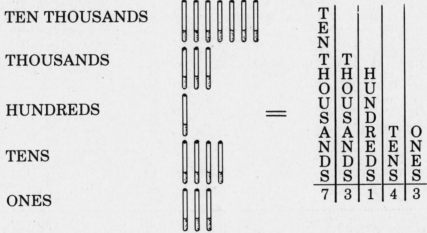

TEN THOUSANDS	THOUSANDS	HUNDREDS	TENS	ONES
7	3	1	4	3

73,143 has 7 ten thousands, 3 thousands, 1 hundred, 4 tens, and 3 ones. The number is read seventy-three thousand, one hundred forty-three.

COMMAS

A comma (,) is used after the number in the thousands place. Commas make the number easier to read. From right to left, put commas after every three places.

Try It (Answers on page 79.)

Write each of the following in numbers. The first one is done for you. Put the comma in the correct place.

1. Thirty-three thousand, four hundred ninety-two = 33,492

2. Sixteen thousand, eight hundred fifty-six = _____

3. Forty-one thousand, nine hundred thirty-seven = _____

4. Twenty-two thousand, two hundred twenty-one = _____

5. Ninety-four thousand, five hundred twelve = _____

6. Fifty-seven thousand, three hundred sixty-eight = _____

7. Eighty-six thousand, five hundred seventy-three = _____

The next place to the left is HUNDRED THOUSANDS. The population of San Juan, Puerto Rico, is 455,421. This number is read: four hundred fifty-five thousand, four hundred twenty-one. The number 455,421 has 4 hundred thousands, 5 ten thousands, 5 thousands, 4 hundreds, 2 tens, and 1 one.

Try It (Answers on page 79.)

Write these numbers in words.

1. 648,927 _____

2. 295,631 _____

3. 324,768 _____

4. 419,253 _____

5. 163,647 _____

The next place to the left is MILLIONS. The atlas says there are 7,781,984 people living in New York City. This number is read: seven million, seven hundred eighty-one thousand, nine hundred eighty-four. The number 7,781,984 has 7 millions, 7 hundred thousands, 8 ten thousands, 1 thousand, 9 hundreds, 8 tens, and 4 ones.

Try It (Answers on page 79.)

Write each of the following in numbers. Put a comma in the correct place.

1. three million, four hundred eighteen thousand, six hundred twenty-three = _____

2. five million, two hundred forty-five thousand, six hundred fifty-one = _____

3. six million, six hundred sixty-one thousand, two hundred seventy-seven = _____

4. four million, one hundred twenty-four thousand, three hundred thirty-eight = _____

5. two million, two hundred twenty-one thousand, eight hundred twelve = _____

6. nine million, three hundred eleven thousand, five hundred forty-three = _____

PUTTING ZERO IN ITS PLACE

A zero is worth nothing. But it can make the difference between having $22 and having $202. 202 has 2 hundreds, 0 tens, and 2 ones. 22 has 2 tens and 2 ones.

In the number 202, zero acts as a place holder. This place holder lets you write hundreds without having any tens.

Try It (Answers on page 79.)

Write each of the following in numbers. Put the comma in the correct place.

1. Five thousand, sixteen =

2. Eleven thousand, one =

3. Fifty thousand =

4. Nine hundred six thousand, forty-two =

5. Forty-three thousand, fourteen =

6. One million =

ROUNDING OFF WHOLE NUMBERS

People do not always use exact numbers. For example, 19,989 people went to a soccer game last Saturday. The sports page in the newspaper reported that "about 20,000 people went to the game." The sports writer <u>rounded off</u> the number 19,989 to the nearest thousand.

To round off a number, follow these steps:

Step 1—Look at the number place you want to round off.

Step 2—Look at the number place to the right of the place you want to round off.

Is the number less than 5? If so, change it and all the numbers to the right of it to zeros.

Is the number 5 or more? If so, increase the number to the left by 1. Then change all the numbers to the right to zeros.

Look at these examples.

Round off 432 to hundreds.

Step 1—4 is in the hundreds place.

Step 2—The number to the right of the 4 is 3. 3 is less than 5. Leave the hundreds number as 4. Change the rest of the numbers to zeros.

432 rounded off to hundreds is 400.

Round off 16,890 to thousands.

Step 1—6 is in the thousands place.

Step 2—The number to the right of the 6 is 8. 8 is more than 5. Increase the thousands number by 1. Change the rest of the numbers to zeros.

16,890 rounded off to thousands is 17,000.

Try It (Answers on page 79.)

1. 52,907 rounded off to thousands is _____

2. 549 rounded off to hundreds is _____

3. 482,580 rounded off to ten thousands is _____

4. 482,580 rounded off to hundred thousands is _____

5. 91,999 rounded off to thousands is _____

3. ADDITION

Adding is quicker than counting. You have 2 pennies in one pile and 3 pennies in another pile. How many do you have altogether?

You can count:

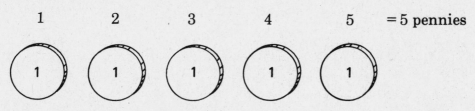

Or you can add:

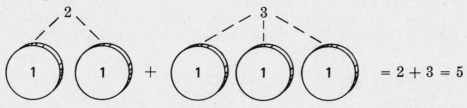

The sign for <u>addition</u> is +. This is called a plus sign. This sign tells you to add the numbers. The answer is called the SUM or TOTAL.

Try It (Answers on page 79.)

Add these numbers.

1.   3
    + 1
	2. 4
	    + 5
		3. 7
		    + 1
			4. 4
			    + 2
				5. 6
				    + 2

6.   1
    + 4
	7. 3
	    + 5
		8. 2
		    + 2
			9. 1
			    + 8
				10. 5
				    + 2

Because people write sentences across the page, they sometimes write addition problems that way. But it is easier to add figures in a straight line going down.

You read: 6 + 3 =   You add: $\begin{array}{r} 6 \\ + 3 \\ \hline \end{array}$

Try It (Answers on page 80.)

Write these problems in a line going down. Then add them and write the sum.

1. $6 + 3 =$ 2. $1 + 5 =$ 3. $2 + 3 =$ 4. $7 + 2 =$

5. $3 + 4 =$ 6. $3 + 3 =$ 7. $2 + 1 =$ 8. $1 + 8 =$

ADDING TENS NUMBERS

Tens numbers are added one column at a time. Look at this example.

		Step 1		Step 2	
	34	Tens	Ones	Tens	Ones
	+ 53	3	4	3	4
		+ 5	3	+ 5	3
			7	8	7

Step 1—Add the numbers in the ones column: $4 + 3 = 7$.

Step 2—Add the numbers in the tens column: $3 + 5 = 8$.

Try It (Answers on page 80.)

1.   24   2.   35   3.   24   4.   28   5.   46
    + 23   + 33   + 61   + 71   + 31

6.   73   7.   11   8.   14   9.   27   10.   62
    + 10   + 43   + 15   + 11   + 27

ADDING HUNDREDS NUMBERS

Hundreds numbers are also added one column at a time. First, add the ones column. Next, add the tens column. Then add the hundreds column. Look at this example:

	Step 1	Step 2	Step 3

245
+ 112
	H T O
	2 4 5
	+ 1 1 2
	    7
		H T O
		2 4 5
		+ 1 1 2
		  5 7
			H T O
			2 4 5
			+ 1 1 2
			3 5 7

Step 1—Add the ones column: $5 + 2 = 7$.

Step 2—Add the tens column: $4 + 1 = 5$.

Step 3—Add the hundreds column: $2 + 1 = 3$.

Try It (Answers on page 80.)

1. 837 + 142	2. 312 + 473	3. 531 + 427	4. 206 + 691	5. 426 + 351
6. 171 + 528	7. 230 + 421	8. 718 + 141	9. 602 + 101	10. 235 + 234

EMPTY COLUMNS

Sometimes, numbers having tens places are added to numbers having only ones places. Or, numbers having hundreds places are added to numbers having only tens or ones places. These are added the same as other problems. When a place is empty, there is nothing to add. You can put a zero in that place to keep the columns straight.

	Step 1	Step 2	Step 3

546
+ 31
	H T O
	5 4 6
	+ 3 1
	    7
		H T O
		5 4 6
		+ 3 1
		  7 7
			H T O
			5 4 6
			+ 3 1
			5 7 7

Step 1—Add the ones column: $6 + 1 = 7$.

Step 2—Add the tens column: $4 + 3 = 7$.

Step 3—Add the hundreds column. There is no number under the five. You can think of this place as zero. $5 + 0 = 5$.

Try It (Answers on page 80.)

1.	42 + 6	2.	307 + 81	3.	621 + 65	4.	151 + 38	5.	223 + 56
6.	47 + 2	7.	485 + 10	8.	52 + 826	9.	266 + 33	10.	813 + 85

ADDING LONGER COLUMNS

You can now add two numbers at a time. Look at the example below. There are three numbers to add. Add the first two numbers. Then add this total to the third number.

$$
\begin{array}{c}
1 \\
3 \\
+\,2 \\
\hline
6
\end{array}
\quad
\begin{array}{l}
1 + 3 = 4 \\
\quad\dots\dots 2
\end{array}
\quad
4 + 2 = 6
$$

Step 1—Add: $1 + 3 = 4$.

Step 2—Add: $4 + 2 = 6$.

Try It (Answers on page 80.)

1.	5 3 + 1	2.	2 5 + 1	3.	1 6 + 1	4.	3 0 + 4	5.	4 1 + 3	6.	1 0 + 0

7. $1 + 1 + 3 =$ _____   10. $2 + 0 + 5 =$ _____

8. $6 + 2 + 1 =$ _____   11. $3 + 3 + 0 =$ _____

9. $2 + 4 + 2 =$ _____   12. $4 + 1 + 2 =$ _____

To add columns of more than three numbers, add two numbers at a time. Look at the example below:

$$
\begin{array}{c}
3 \\
2 \\
1 \\
+\,2 \\
\hline
8
\end{array}
\quad
\begin{array}{l}
3 + 2 = 5 \\
\quad\dots\dots 1 \\
\quad\dots\dots\dots\dots 2
\end{array}
\quad
\begin{array}{l}
5 + 1 = 6 \\
\quad 6 + 2 = 8
\end{array}
$$

Step 1—Add: $3 + 2 = 5$.

Step 2—Add: $5 + 1 = 6$.

Step 3—Add: $6 + 2 = 8$.

You can add the numbers in any order.

Try It (Answers on page 80.)

1.	2.	3.	4.	5.
1	0	5	3	1
3	7	2	2	2
2	2	1	3	1
+ 2	+ 0	+ 1	+ 0	+ 2

6.	7.	8.	9.	10.
4	5	3	2	3
1	1	0	1	4
2	1	3	4	0
1	1	0	1	1
+ 1	+ 1	+ 3	+ 0	+ 1

Look at the example below. You must add numbers having tens and ones columns. First add the ones column. Then add the tens column.

		Step 1	Step 2
		T|O	T|O
13		1|3	1|3
23		2|3	2|3
+ 51		+5|1	+5|1
		|7	8|7

Step 1—Add the ones column: $3 + 3 = 6$, $6 + 1 = 7$.

Step 2—Add the tens column: $1 + 2 = 3$, $3 + 5 = 8$.

You add the same way no matter how large the numbers are. Always add the ones column first. Then keep moving one column at a time going to the left. Look at this example:

	Th|H|T|O	Th|H|T|O	Th|H|T|O	Th|H|T|O
1,143	1|1|4|3	1|1|4|3	1|1|4|3	1|1|4|3
2,021	2|0|2|1	2|0|2|1	2|0|2|1	2|0|2|1
+ 4,411	+4|4|1|1	+4|4|1|1	+4|4|1|1	+4|4|1|1
	|||5	||7|5	|5|7|5	7|5|7|5

Try It (Answers on page 80.)

1.	41	2.	32	3.	12	4.	51	5.	16
	32		41		23		14		30
	+ 23		+ 12		+ 10		+ 22		+ 20

6.	33	7.	43	8.	15	9.	24	10.	37
	11		26		12		21		21
	+ 24		+ 30		+ 10		+ 42		+ 21

11.	432	12.	600	13.	365	14.	217	15.	114
	123		12		421		32		200
	+ 301		+ 304		+ 100		+ 20		+ 4

16.	1,314	17.	3,016	18.	5,212	19.	7,000
	2,132		4,620		521		81
	+ 4,351		+ 2,203		+ 52		+ 408

WORKING WITH WORD PROBLEMS

Read the Steps for Solving Word Problems on page xii again. Then work these problems. Find the correct answer.

Here are some important clue words that will help you solve addition word problems. These clue words usually mean you must add.

1. **How much** or **how many**—The question might ask, "How much money was spent?" Or, "How many drinks were served?"

2. **Total** or **sum**—The question might ask, "What is the total amount of money spent?"

3. **More**—The question might ask, "If he had 5 dollars more this week, how much did he have altogether?"

4. **Added to**—The question might say, "Then 5 dollars was added to the amount he had."

5. **Increased**—The question might say, "The bus fare was increased 15 cents."

6. **Altogether**—The question might ask, "How many are there altogether?"

Now, work these word problems.

1. At the beginning of the week, Ana had two record albums. On Tuesday, she bought three more. On Thursday, she bought another. How many record albums did she have at the end of the week?

(1) 5   (2) 6   (3) 7   (4) 4

2. Carl bought two cans of soda, two candy bars, three packs of gum, and a newspaper. What is the total number of items he bought?

(1) 5   (2) 4   (3) 7   (4) 8

3. It is two blocks from Mr. Solano's apartment to the grocery store, three more blocks to the bowling alley and three more blocks to the bank. How many blocks is it from his apartment to the bank?

(1) 3   (2) 5   (3) 8   (4) 9

4. Peter worked 37 hours last week and 12 hours overtime. How many hours did he work?

(1) 47   (2) 48   (3) 49   (4) 39

5. On 59th Street in New York City, 11 cars were towed away on Monday. No cars were towed away on Tuesday. Twenty-one cars were towed away on Wednesday. Five cars were towed away on Thursday, and 10 cars were towed away on Friday. How many cars were towed away all week?

(1) 47   (2) 92   (3) 37   (4) 36

6. Ten people on the block decided to form a neighborhood council. In addition to these 10 people, 4 more people came to the first meeting, 12 new people came to the second meeting, and 3 new people came to the third meeting. What was the total number of people at the third meeting?

(1) 29   (2) 25   (3) 27   (4) 30

7. Two years ago, there were 112 children enrolled in a city's day care program. Last year, 231 more children enrolled. This year 315 more children enrolled. What is the total number of children enrolled in the city's day care program?

(1) 668   (2) 556   (3) 658   (4) 657

8. At a factory, 421 people work the first shift, 355 people work the second shift, and 112 people work the third shift. What is the total number of people working at the factory?

(1) 887   (2) 787   (3) 888   (4) 878

9. During the month, a coffee shop served 113 breakfasts the first week, 102 the second week, 141 the third week, and 133 the fourth week. How many breakfasts were served that month?
(1) 479   (2) 478   (3) 497   (4) 489

10. The Andretti family pays 6,324 dollars a year for the mortgage and taxes on their house. They spend 6,260 dollars a year on food and clothing. What is the total amount of money the Andrettis spend each year for food, clothing, and shelter?
(1) $12,494   (2) $12,548   (3) $12,584   (4) $12,590

Check your answers on page 81.

4. MORE ON ADDITION

Look at the example. Remember that 10 ones is worth 1 ten. 8 ones added to 7 ones is 15 ones. This answer is the same as 10 ones and 5 ones, or 1 ten and 5 ones.

In addition problems, whenever there are 10 or more in one place column, the 1 is <u>carried</u> to the next column. 8 + 7 = 15. Carry the 1 to the next column.

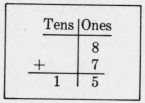

Try It (Answers on page 81.)

Add. The answers will be 10 or more.

1.	5 + 5	2.	9 + 8	3.	6 + 7	4.	4 + 8	5.	6 + 5

6.	9 + 9	7.	7 + 8	8.	6 + 9	9.	5 + 8	10.	8 + 8

In the examples below, the answers in the shaded column are 10 or more. The 1 is <u>carried</u> to the next column.

H	T	O
	6	3
+	7	2
1	3	5

Th	H	T	O
	6	2	6
+	4	4	1
1	0	6	7

Try It (Answers on page 81.)

1.	71 + 45	2.	48 + 81	3.	94 + 94	4.	36 + 92	5.	85 + 74

6.	723 + 655	7.	962 + 704	8.	312 + 734	9.	540 + 603	10.	858 + 631

19

CARRYING TO THE TENS COLUMN

Look at the example below. The 10 ones are carried as 1 ten.

	Step 1	Step 2	Step 3

	Tens	Ones	Tens	Ones	Tens	Ones
38	3	8	$\overset{1}{3}$	8	$\overset{1}{3}$	8
+ 47	+ 4	7	+ 4	7	+ 4	7
		15		5	8	5

Step 1—Add the ones column: $8 + 7 = 15$. The 1 in 15 is one ten.

Step 2—Carry the 1 from the 15 as ten and place it in the tens column. Leave the 5.

Step 3—Add the 1 to the tens column: $1 + 3 + 4 = 8$.

Try It (Answers on page 81.)

Add. Some of the problems have been started for you.

1.
$$\overset{1}{}24$$
$$+ 38$$
$$\overline{2}$$

2.
$$\overset{1}{}46$$
$$+ 35$$
$$\overline{1}$$

3.
$$\overset{1}{}75$$
$$+ 15$$
$$\overline{0}$$

4.
$$\overset{1}{}38$$
$$+ 8$$
$$\overline{6}$$

5.
$$\overset{1}{}69$$
$$+ 25$$
$$\overline{4}$$

6.
$$79$$
$$+ 16$$

7.
$$28$$
$$+ 9$$

8.
$$43$$
$$+ 7$$

9.
$$56$$
$$+ 25$$

10.
$$47$$
$$+ 17$$

11.
$$14$$
$$+ 69$$

12.
$$44$$
$$+ 8$$

13.
$$77$$
$$+ 19$$

14.
$$54$$
$$+ 29$$

15.
$$26$$
$$+ 37$$

16.
$$59$$
$$+ 79$$

17.
$$77$$
$$+ 66$$

18.
$$75$$
$$+ 86$$

19.
$$64$$
$$+ 46$$

20.
$$97$$
$$+ 3$$

CARRYING TO THE HUNDREDS COLUMN

Look at the example on the next page. The 10 tens are carried as 1 hundred and added to the hundreds column.

		Step 1	Step 2	Step 3

	Step 1	Step 2	Step 3
	H T O	H T O	H T O

483
+ 361
	4 8 3
	+ 3 6 1
	    4
		¹
		4 8 3
		+ 3 6 1
		  4 4
			¹
			4 8 3
			+ 3 6 1
			8 4 4

Step 1—Add the ones column: 3 + 1 = 4.

Step 2—Add the tens column: 8 + 6 = 14. Carry the 10 tens as 1 hundred.

Step 3—Add the 1 to the hundreds column: 1 + 4 + 3 = 8.

Try It (Answers on page 82.)

1.   652
  + 173
	2. 478
	  + 261
		3. 324
		  + 492
			4. 286
			  + 243
				5. 565
				  + 392

6.   440
  + 260
	7. 192
	  + 192
		8. 766
		  + 173
			9. 327
			  + 491
				10. 458
				  + 451

11.   697
  + 782
	12. 531
	  + 896
		13. 770
		  + 459
			14. 984
			  + 984
				15. 856
				  + 550

CARRYING TO THE THOUSANDS COLUMN

Look at this example. It shows 10 hundreds being carried as 1 thousand and added to the thousands column.

	Step 1	Step 2	Step 3	Step 4
	Th H T O	Th H T O	Th H T O	Th H T O

4,732
+ 3,624
	4 7 3 2
	+ 3 6 2 4
	      6
		4 7 3 2
		+ 3 6 2 4
		    5 6
			4 7 3 2
			+ 3 6 2 4
			  3 5 6
				¹
				4 7 3 2
				+ 3 6 2 4
				8 3 5 6

Step 1—Add the ones column: 2 + 4 = 6.

Step 2—Add the tens column: 3 + 2 = 5.

Step 3—Add the hundreds column: 7 + 6 = 13. Carry the 1 from the 13 and place it in the next column.

Step 4—Add the 1 to the thousands column: 1 + 4 + 3 = 8.

Try It (Answers on page 82.)

1. 2,673
   + 3,622

2. 5,961
   + 2,324

3. 6,850
   + 742

4. 2,968
   + 1,431

5. 3,624
   + 4,434

6. 6,982
   + 2,817

7. 2,326
   + 4,713

8. 3,604
   + 2,772

9. 1,811
   + 7,311

10. 4,642
    + 2,547

CARRYING NUMBERS IN SEVERAL COLUMNS

Look at the example below. A 1 has to be carried in three columns.

	7,684		Th	H	T	O		Th	H	T	O		Th	H	T	O		Th	H	T	O
+	469		7	6	8¹	4		7	6¹	8¹	4		¹7	6¹	8	4		¹7	6	8	4
		+		4	6	9	+		4	6	9	+		4	6	9	+		4	6	9
						3				5	3			1	5	3		8	1	5	3

Sometimes the numbers in a column add up to more than one group of 10. You can carry any number of groups. Look at the example below.

	1,632		Th	H	T	O		Th	H	T	O		Th	H	T	O
	798		1	6	3²	2		1	6	3²	2			1	6	3²	2
	429			7	9	8			7	9	8				7	9	8
+	199	+		4	2	9	+		4	2	9	+			4	2	9
				1	9	9			1	9	9				1	9	9
						8				5	8			3	0	5	8

Try It (Answers on page 82.)

Each problem has at least one number that must be carried. In some problems, the number to be carried is more than 1.

1. 6,406
   + 2,718

2. 5,384
   + 1,337

3. 3,875
   + 4,546

4. 2,001
   + 1,999

5. 5,784
   + 6,347

6. 7,614
   + 5,786

7. 2,864
   + 7,539

8. 6,302
   + 5,818

9.	4,312 3,041 + 2,864	10.	6,745 8,340 + 1,462	11.	9,837 159 + 3,566	12.	934 6,289 + 8,304
13.	3,241 304 4,653 + 41	14.	95 2,637 2,080 + 5,745	15.	478 963 8,109 + 71	16.	147 8,563 35 + 8,687
17.	8,387 8,289 3,137 + 1,236	18.	2,307 + 8,923	19.	483 + 9,617	20.	23 847 2,483 + 6

WORKING WITH WORD PROBLEMS

Read the Steps for Solving Word Problems on page xii again. Then work these problems. Find the correct answer.

1. Lou was given seven speeding tickets last year. He has four speeding tickets so far this year. What is the total number of speeding tickets Lou has received?
   (1) 9   (2) 10   (3) 11   (4) 12

2. At the first stop, seven people got on the bus. At the second stop, four people got on. At the next stop, three people got on. At the fourth stop, seven people got on. How many people were on the bus after the fourth stop?
   (1) 20   (2) 21   (3) 17   (4) 22

3. Eighth Street has three apartment buildings. The first building has 36 families. The second building has 31 families. The third building has 40 families. How many families live in apartment buildings on Eighth Street?
   (1) 107   (2) 97   (3) 77   (4) 67

4. A job training program enrolled 166 secretaries, 242 mechanics, 135 practical nurses, and 154 machine operators. What is the total number of persons enrolled in the job training program?
   (1) 587   (2) 697   (3) 687   (4) 597

5. West Side School employs 92 teachers' aides. East Side School employs 126 teachers' aides. South Side School employs 85 teachers' aides. Central School employs 102 teachers' aides. How many teachers' aides are employed by these four schools?
   (1) 311   (2) 405   (3) 511   (4) 521

6. Clark Clinic received 145 patients on Monday, 273 patients on Tuesday, 198 patients on Wednesday, 205 patients on Thursday, and 210 patients on Friday. How many patients visited the clinic that week?
   (1) 812   (2) 832   (3) 1,031   (4) 1,131

7. In Central City, 1,263 people registered for unemployment benefits in September, 1,725 more people registered in October, and 876 people registered in November. How many people registered for unemployment benefits in these three months?
   (1) 3,864   (2) 3,764   (3) 3,854   (4) 2,754

8. A labor union has locals in five cities. The first city has 947 members in its local. The second city has 1,093 members. The third city has 756 members. The fourth city has 1,100 members. The last city has 1,000 members. How many members are there in all five locals together?
   (1) 4,869   (2) 4,896   (3) 4,796   (4) 4,886

9. A voter registration drive registered 2,550 people in 1986, 1,437 people in 1987, 880 in 1988, and 643 in 1989. How many voters were registered in these four years?
   (1) 3,200   (2) 5,310   (3) 5,410   (4) 5,510

10. Exactly 40,963 people attended the first game of the world series, 37,461 people went to the second game, and 40,000 people went to the third game. If 34,312 people went to the fourth game and 50,023 people saw the fifth and final game, what was the total attendance at the world series?
    (1) 182,759   (2) 191,769   (3) 200,759   (4) 202,759

Check your answers on page 82.

5. SUBTRACTION

You have $5. You give your brother $2. How much money do you have left? To find out, you subtract.

$$\begin{array}{r} 5 \text{ dollars} \\ - \ 2 \text{ dollars} \\ \hline 3 \text{ dollars} \end{array}$$

The number you gave away is placed <u>under</u> the amount you had.

Subtraction is "taking away" one number from another. You "took away" $2 from $5.

The sign for subtraction is called a minus sign. It looks like a dash (−). Whenever you see this sign, you know that you must subtract. The answer to a subtraction problem is called the DIFFERENCE.

CHECKING SUBTRACTION

There is a way to check subtraction. Add the answer and the number taken away. If the answer is correct, these two numbers will add up to the number subtracted from. Look at the example below:

$$\text{Problem:} \quad \begin{array}{r} 5 \text{ dollars} \\ - \ 2 \text{ dollars} \\ \hline 3 \text{ dollars} \end{array} \qquad \text{Check:} \quad \begin{array}{r} 2 \text{ dollars} \\ + \ 3 \text{ dollars} \\ \hline 5 \text{ dollars} \end{array}$$

Try It (Answers on page 83.)

Subtract. Check by adding.

1. $\begin{array}{r} 3 \\ - \ 2 \\ \hline \end{array}$
2. $\begin{array}{r} 7 \\ - \ 5 \\ \hline \end{array}$
3. $\begin{array}{r} 5 \\ - \ 1 \\ \hline \end{array}$
4. $\begin{array}{r} 9 \\ - \ 3 \\ \hline \end{array}$
5. $\begin{array}{r} 6 \\ - \ 4 \\ \hline \end{array}$
6. $\begin{array}{r} 4 \\ - \ 3 \\ \hline \end{array}$

7. $\begin{array}{r} 8 \\ - \ 4 \\ \hline \end{array}$
8. $\begin{array}{r} 2 \\ - \ 1 \\ \hline \end{array}$
9. $\begin{array}{r} 7 \\ - \ 4 \\ \hline \end{array}$
10. $\begin{array}{r} 4 \\ - \ 2 \\ \hline \end{array}$
11. $\begin{array}{r} 6 \\ - \ 3 \\ \hline \end{array}$
12. $\begin{array}{r} 5 \\ - \ 3 \\ \hline \end{array}$

PROBLEMS WRITTEN ACROSS

Sometimes subtraction problems are written across the page. The number you subtract, or take away, is always written <u>after</u> the number you subtract from. Place it <u>under</u> the number you subtract from when you work the problem.

You read: $6 - 4 = 2$ You subtract: $\begin{array}{r} 6 \\ -\ 4 \\ \hline 2 \end{array}$

Try It (Answers on page 83.)

Write these problems in a line going down. Subtract them. Check by adding.

1. $7 - 3 =$ 2. $9 - 4 =$ 3. $3 - 1 =$ 4. $6 - 5 =$

5. $8 - 3 =$ 6. $5 - 2 =$ 7. $9 - 7 =$ 8. $8 - 2 =$

SUBTRACTING TENS NUMBERS

To subtract numbers that have tens and ones places, begin by subtracting in the ones column. Then subtract the tens column. Look at the example below:

		Step 1		Step 2	
		Tens	Ones	Tens	Ones
2 6		2	6	2	6
− 1 4		− 1	4	− 1	4
			2	1	2

Step 1—Subtract the ones column: $6 - 4 = 2$.

Step 2—Subtract the tens column: $2 - 1 = 1$.

Check your answer by adding the number you took away and your answer. Look at the example below:

Problem:	T	O	Check:	T	O
	2	6		1	4
	− 1	4		+ 1	2
	1	2		2	6

Try It (Answers on page 83.)

Subtract. Check by adding.

1. 39 − 26	2. 65 − 51	3. 47 − 12	4. 55 − 33	5. 73 − 52

6. 89 − 54	7. 35 − 21	8. 98 − 64	9. 77 − 46	10. 67 − 22

11. 25 − 12	12. 74 − 50	13. 82 − 31	14. 49 − 17	15. 37 − 11

SUBTRACTING LARGER NUMBERS

Once you have learned to subtract numbers having ones and tens columns, you can subtract larger numbers. Look at the example below.

	Step 1	Step 2	Step 3

263
− 142
	H T O
	2 6 3
	− 1 4 2
	    1
		H T O
		2 6 3
		− 1 4 2
		  2 1
			H T O
			2 6 3
			− 1 4 2
			1 2 1

Step 1—Subtract the ones column: $3 - 2 = 1$.

Step 2—Subtract the tens column: $6 - 4 = 2$.

Step 3—Subtract the hundreds column: $2 - 1 = 1$.

Problem:	263 − 142 121	Check:	142 + 121 263

Numbers having thousands and ten thousands places are worked out by subtracting each column separately. You work from right to left. Be sure your problems are written neatly so that you do not subtract numbers from the wrong columns. Check your work by adding.

Try It (Answers on page 83.)

Subtract. Check by adding.

| 1. | 496 − 284 | 2. | 548 − 113 | 3. | 838 − 705 | 4. | 685 − 342 | 5. | 727 − 214 |

| 6. | 2,998 − 1,756 | 7. | 5,742 − 3,311 | 8. | 7,306 − 5,105 | 9. | 8,656 − 6,531 | 10. | 7,439 − 6,417 |

| 11. | 97,563 − 62,452 | 12. | 93,783 − 21,422 | 13. | 84,865 − 32,555 | 14. | 74,607 − 64,403 | 15. | 39,829 − 16,416 |

ZERO HAS A PLACE

You have $35 and spend $25. To find out how much you have left, you must subtract. Look at the example below:

$$\begin{array}{r} 35 \text{ dollars} \\ -\ 25 \text{ dollars} \\ \hline 10 \text{ dollars} \end{array}$$

The zero must be written down in the answer. It acts as a place holder. Would you rather have $1 or $10 left? The zero tells you there is $10 left, not $1.

Look at the examples below. They show how zero acts as a place holder.

$$\begin{array}{r} 96 \\ -\ 26 \\ \hline 70 \end{array} \qquad \begin{array}{r} 953 \\ -\ 651 \\ \hline 302 \end{array} \qquad \begin{array}{r} 4,602 \\ -\ 1,601 \\ \hline 3,001 \end{array}$$

Try It (Answers on page 83.)

Subtract. Check by adding.

| 1. | 23 − 13 | 2. | 213 − 113 | 3. | 788 − 681 | 4. | 548 − 138 | 5. | 605 − 203 |

| 6. | 3,889 − 1,865 | 7. | 9,694 − 2,692 | 8. | 5,000 − 3,000 | 9. | 6,572 − 4,531 | 10. | 3,224 − 1,214 |

EMPTY COLUMNS

Look at the examples below.

$$
\begin{array}{r} 427 \\ -\ 14 \\ \hline 413 \end{array}
\qquad
\begin{array}{r} 8,648 \\ -\ \ \ 27 \\ \hline 8,621 \end{array}
$$

In the first example, there is no number to subtract from the 4 hundreds. Think $4 - 0 = 4$. Bring the 4 down in the answer.

In the second example, there are two places with nothing to subtract. $6 - 0 = 6$. $8 - 0 = 8$.

Try It (Answers on page 83.)

Subtract. Check by adding.

1.	2.	3.	4.	5.
37 − 4	89 − 6	25 − 2	185 − 22	868 − 7

6.	7.	8.	9.	10.
2,987 − 243	7,647 − 16	8,599 − 7	3,536 − 5	1,035 − 3

WORKING WITH WORD PROBLEMS

Read the Steps for Solving Word Problems on page xii again. Then work these problems. Find the correct answer.

Here are some important subtraction clue words. They can help you solve word problems in subtraction.

1. **Less**—The question might ask, "How much less was spent?"

2. **Fewer**—The question might ask, "How many fewer cars were sold?"

3. **Difference**—The question might ask, "What is the difference in the prices?"

4. **Left** or **Remains**—The question might ask, "How much is left?" Or, "How much remains?"

5. **Decreased**—The question might ask, "By how much was the total decreased?"

6. **How many more**—A problem may give you two numbers. It will ask how much larger one number is. Subtract the smaller number from the larger number. The question might ask, "If I got 15 letters today and 5 letters yesterday, how many more did I get today?"

Now work these word problems.

1. Clara had nine dollars when she went shopping. By the end of the day, she had spent a total of five dollars. How much does she have left?
   (1) $5   (2) $14   (3) $3   (4) $4

2. Mrs. Kaplan found a dress she wanted. It cost 50 dollars in the store. It would cost her 26 dollars to sew at home. How much can she save by sewing the dress at home?
   (1) $34   (2) $24   (3) $22   (4) $14

3. Harlem School graduated a total of 193 students this year. This is 51 more students than were graduated last year. How many students were graduated last year?
   (1) 142   (2) 244   (3) 144   (4) 92

4. Last year Carlos weighed 147 pounds. He lost 15 pounds last month. How many pounds does he weigh now?
   (1) 132   (2) 142   (3) 162   (4) 167

5. Last week Helen picked 356 bushels of tomatoes. This week she picked 466 bushels. How many more bushels of tomatoes did she pick this week?
   (1) 59   (2) 151   (3) 149   (4) 110

6. Phil's car goes 9 miles on one gallon of gas. His wife's car goes 19 miles on a gallon. How many more miles does her car go on a gallon?
   (1) 10   (2) 14   (3) 32   (4) 18

7. The basketball team won 30 out of 60 games this season. How many games did they lose?
   (1) 20   (2) 30   (3) 31   (4) 39

8. Last year 56,430 people read *True Love* magazine. The same year, 97,530 people read *Sports* magazine. How many more people read *Sports* than *True Love*?
   (1) 4,110   (2) 41,000   (3) 41,010   (4) 41,100

9. Plotnik Auto Works laid off 42 people last year. The company says it has laid off 12 fewer people this year. How many people were laid off this year?
   (1) 34   (2) 32   (3) 30   (4) 3

10. The manager of Slats Clothing Store is unhappy about decreasing sales. Last week he had sales of $1,908. This week he had sales of $805. By how much did his sales decrease?
    (1) $103   (2) $113   (3) $1,003 (4) $1,103

Check your answers on page 84.

6. MORE ON SUBTRACTION

Sometimes you have to regroup numbers to subtract. Look at the example below:

	Step 1				Step 2		
	Tens	Ones			Tens	Ones	
1 6 − 8	1 −	6 8	You cannot subtract.		1 6 − 8	You can subtract.

Step 1—You cannot subtract 8 from 6. Borrow 1 ten from the tens column as 10 ones: 1 ten = 10 ones. The 1 ten is regrouped with the 6 ones to make 16 ones.

Step 2—Subtract 8 from 16: $16 - 8 = 8$.

$$\begin{array}{r} \text{Problem:} \quad 16 \\ -\ 8 \\ \hline 8 \end{array} \qquad \begin{array}{r} \text{Check:} \quad 8 \\ +\ 8 \\ \hline 16 \end{array}$$

Try It (Answers on page 84.)

Subtract. Check by adding.

1. $\begin{array}{r} 18 \\ -\ 9 \\ \hline \end{array}$ 2. $\begin{array}{r} 15 \\ -\ 8 \\ \hline \end{array}$ 3. $\begin{array}{r} 13 \\ -\ 4 \\ \hline \end{array}$ 4. $\begin{array}{r} 17 \\ -\ 9 \\ \hline \end{array}$ 5. $\begin{array}{r} 11 \\ -\ 3 \\ \hline \end{array}$

6. $\begin{array}{r} 12 \\ -\ 6 \\ \hline \end{array}$ 7. $\begin{array}{r} 14 \\ -\ 6 \\ \hline \end{array}$ 8. $\begin{array}{r} 16 \\ -\ 7 \\ \hline \end{array}$ 9. $\begin{array}{r} 15 \\ -\ 6 \\ \hline \end{array}$ 10. $\begin{array}{r} 13 \\ -\ 6 \\ \hline \end{array}$

BORROWING FROM THE TENS COLUMN

Look at the example on the next page. One ten from the tens column is borrowed as 10 ones and placed in the ones column.

	Step 1		Step 2		Step 3	
	Tens	Ones	Tens	Ones	Tens	Ones
36 − 18	$\overset{2}{\cancel{3}}$ − 1	$\overset{16}{\cancel{6}}$ 8	$\overset{2}{\cancel{3}}$ − 1	$\overset{16}{\cancel{6}}$ 8 8	$\overset{2}{\cancel{3}}$ − 1 1	$\overset{16}{\cancel{6}}$ 8 8

Step 1—You cannot subtract 8 from 6. Borrow 1 ten from the tens column as 10 ones. Add it to the 6 in the ones column. Now there is 16 in the ones column.

Step 2—Subtract: $16 - 8 = 8$.

Step 3—Subtract. Remember, you borrowed only 1 ten. $2 - 1 = 1$.

$$
\begin{array}{ll}
\text{Problem:} & \begin{array}{r} 36 \\ -18 \\ \hline 18 \end{array} \qquad
\text{Check:} & \begin{array}{r} \overset{1}{1}8 \\ +18 \\ \hline 36 \end{array}
\end{array}
$$

Try It (Answers on page 84.)

Subtract. Some of the problems have been started for you. Check by adding.

1. $\begin{array}{r} \overset{3\ 13}{\cancel{4}\cancel{3}} \\ -19 \\ \hline \end{array}$ 2. $\begin{array}{r} \overset{7\ 16}{\cancel{8}\cancel{6}} \\ -38 \\ \hline \end{array}$ 3. $\begin{array}{r} \overset{6\ 14}{\cancel{7}\cancel{4}} \\ -55 \\ \hline \end{array}$ 4. $\begin{array}{r} \overset{4\ 12}{\cancel{5}\cancel{2}} \\ -25 \\ \hline \end{array}$ 5. $\begin{array}{r} \overset{4\ 17}{\cancel{5}\cancel{7}} \\ -49 \\ \hline \end{array}$

6. $\begin{array}{r} 75 \\ -37 \\ \hline \end{array}$ 7. $\begin{array}{r} 33 \\ -18 \\ \hline \end{array}$ 8. $\begin{array}{r} 95 \\ -66 \\ \hline \end{array}$ 9. $\begin{array}{r} 54 \\ -37 \\ \hline \end{array}$ 10. $\begin{array}{r} 48 \\ -29 \\ \hline \end{array}$

11. $\begin{array}{r} 53 \\ -15 \\ \hline \end{array}$ 12. $\begin{array}{r} 54 \\ -45 \\ \hline \end{array}$ 13. $\begin{array}{r} 82 \\ -45 \\ \hline \end{array}$ 14. $\begin{array}{r} 66 \\ -37 \\ \hline \end{array}$ 15. $\begin{array}{r} 97 \\ -49 \\ \hline \end{array}$

16. $\begin{array}{r} 35 \\ -18 \\ \hline \end{array}$ 17. $\begin{array}{r} 88 \\ -59 \\ \hline \end{array}$ 18. $\begin{array}{r} 62 \\ -46 \\ \hline \end{array}$ 19. $\begin{array}{r} 95 \\ -56 \\ \hline \end{array}$ 20. $\begin{array}{r} 84 \\ -28 \\ \hline \end{array}$

BORROWING FROM THE HUNDREDS COLUMN

Sometimes you cannot subtract in the tens column. You can borrow 1 hundred from the hundreds column as 10 tens and place it in the tens column.

	Step 1	Step 2	Step 3	Check
436 − 263	436 − 263 3	³¹³ (crossed) 4̶3̶6̶ − 263 73	³¹³ (crossed) 4̶3̶6̶ − 263 173	¹ 263 + 173 436

Step 1—Subtract the ones column: 6 − 3 = 3.

Step 2—You cannot subtract the tens column. Borrow 1 hundred from the hundreds column as 10 tens. Add it to the 3. Now there is 13 in the tens column. Subtract the tens column: 13 − 6 = 7.

Step 3—Subtract the hundreds column: 3 − 2 = 1.

Try It (Answers on page 84.)

Subtract. Check by adding.

1. 967 − 385	2. 642 − 161	3. 535 − 253	4. 784 − 293	5. 427 − 134
6. 625 − 442	7. 348 − 185	8. 737 − 343	9. 563 − 482	10. 486 − 193
11. 645 − 64	12. 352 − 91	13. 168 − 95	14. 227 − 72	15. 433 − 51

BORROWING FROM THE THOUSANDS COLUMN

You have now learned to borrow 1 ten as 10 ones and 1 hundred as 10 tens to solve subtraction problems. You can also borrow 1 thousand from the thousands column as 10 hundreds and place it in the hundreds column. Look at the example below:

	Step 1	Step 2	Check
	Th	H	T	O		Th	H	T	O	
9,473 − 6,522	⁸9̶	¹⁴4̶	7	3		⁸9̶	¹⁴4̶	7	3		¹ 6,522
	− 6	5	2	2		− 6	5	2	2		+ 2,951
		9	5	1		2	9	5	1		9,473

Step 1 shows this problem already worked out to the hundreds place.

Step 1—Borrow 1 thousand from the thousands column as 10 hundreds. Add it to the 4. Subtract: $14 - 5 = 9$.

Step 2—Remember, you borrowed 1 from the thousands column. Subtract the thousands column: $8 - 6 = 2$.

Try It (Answers on page 84.)

Subtract. Check by adding.

1.   8,432
     − 2,511
	2. 4,795
	     − 2,963
		3. 5,546
		     − 1,814
			4. 6,327
			     − 2,613

5.   7,198
     − 3,725
	6. 3,694
	     − 1,973
		7. 9,576
		     − 4,755
			8. 3,377
			     − 1,461

9.   8,684
     −   751
	10. 2,359
	      − 626
		11. 7,168
		      − 753
			12. 9,898
			      − 942

BORROWING FROM MORE THAN ONE COLUMN

Look at the example below. It shows how numbers can be borrowed from more than one column in the same problem.

	Step 1	Step 2	Step 3	Step 4	Check
5,734 − 2,868	5,73⁴ − 2,868 —— 6	5,734 − 2,868 —— 66	5,734 − 2,868 —— 866	5,734 − 2,868 —— 2,866	2,868 + 2,866 —— 5,734

Step 1—Borrow 1 ten from the tens column to subtract the ones column: $14 - 8 = 6$.

Step 2—Borrow 1 hundred from the hundreds column to subtract the tens column: $12 - 6 = 6$.

Step 3—Borrow 1 thousand from the thousands column to subtract the hundreds column: $16 - 8 = 8$.

Step 4—Subtract the thousands column: $4 - 2 = 2$.

Try It (Answers on page 85.)

Subtract. Check by adding.

1.	364	2.	678	3.	522	4.	742	5.	826
	− 188		− 489		− 333		− 175		− 668

6.	2,457	7.	6,655	8.	3,423	9.	5,784	10.	8,694
	− 1,589		− 4,777		− 2,846		− 2,893		− 6,848

11.	7,343	12.	4,521	13.	9,743	14.	6,231	15.	2,652
	− 298		− 618		− 867		− 98		− 87

BORROWING—COLUMNS WITH A ZERO

Sometimes in subtracting you need to borrow and can't. The column you want to borrow from has a zero in it. You can't borrow from a zero. You will have to borrow twice. Look at this example.

$$
\begin{array}{r}
703 \\
-\,169 \\
\end{array}
\qquad
\begin{array}{r}
\overset{6}{\cancel{7}}\,\overset{9}{\cancel{0}}\,\overset{13}{\cancel{3}} \\
-\,1\ 6\ 9 \\
\hline
5\ 3\ 4 \\
\end{array}
$$

Step 1—You cannot subtract 9 from 3. The next column has a zero in it. You cannot borrow from a zero. You will have to go to the next column. Borrow 1 hundred from the 7. $7 - 1 = 6$. Carry the 1 hundred as 10 tens. This makes the zero a 10. Now you can borrow 1 ten as 10 ones. $10 - 1 = 9$. You now have 9 tens and 13 ones. You can subtract.

Step 2—Subtract the ones column: $13 - 9 = 4$.

Step 3—Subtract the tens column: $9 - 6 = 3$.

Step 4—Subtract the hundreds column: $6 - 1 = 5$.

Try It (Answers on page 85.)

Subtract. Check by adding.

1.	802	2.	608	3.	407	4.	306	5.	500
	− 153		− 439		− 268		− 147		− 123

6. 6,503	7. 5,034	8. 7,040	9. 8,065	10. 9,305
− 1,265	− 2,842	− 3,384	− 5,978	− 5,648

TWO ZEROS IN A ROW

Sometimes you must borrow three times. Look at the example below:

$$
\begin{array}{r}
8,006 \\
- 2,629 \\
\hline
\end{array}
\qquad
\begin{array}{r}
\overset{9}{}\overset{9}{} \\
7\ 10\ 10\ 16 \\
8\,0\,0\,6 \\
-\ 2\ 6\ 2\ 9 \\
\hline
5\ 3\ 7\ 7 \\
\end{array}
$$

Step 1—You cannot subtract 9 from 6. You must begin borrowing from the thousands column. Borrow 1 thousand as 10 hundreds. Then borrow 1 hundred as 10 tens. Then borrow 1 ten as 10 ones. Now you can subtract.

Step 2—Subtract the ones column: $16 - 9 = 7$.

Step 3—Subtract the tens column: $9 - 2 = 7$.

Step 4—Subtract the hundreds column: $9 - 6 = 3$.

Step 5—Subtract the thousands column: $7 - 2 = 5$.

Try It (Answers on page 85.)

1. 7,004	2. 2,001	3. 5,006	4. 1,000	5. 3,008
− 4,395	− 1,969	− 2,637	− 679	− 1,729

6. 9,003	7. 12,007	8. 51,002	9. 86,003	10. 43,000
− 998	− 2,638	− 6,075	− 8,527	− 5,023

WORKING WITH WORD PROBLEMS

Read the Steps for Solving Word Problems on page xii again. Then work these problems. Find the correct answer.

1. Mr. Drummond has $11 in his pocket. He wants to spend $5 to go to the movies. How much will he have left?
   (1) $4   (2) $6   (3) $16   (4) $7

2. A portable TV set costs $180 at the Valley TV store. The same TV set costs $165 at the Mountain TV store. How much will you save at the Mountain TV store?
   (1) $35   (2) $45   (3) $25   (4) $15

3. A basketball player threw 38 baskets. He made 29 of them. How many baskets did he miss?
   (1) 8   (2) 9   (3) 19   (4) 28

4. Frank and his wife have a combined yearly salary of $18,421. Frank earns $10,638. How much does his wife earn?
   (1) $8,783   (2) $8,893   (3) $7,783   (4) $7,893

5. At Christmas time, the Post Office hired 145 extra people. A total of 540 people worked there at Christmas. How many people worked there before Christmas?
   (1) 405   (2) 395   (3) 495   (4) 245

6. Last month a school lunch program served 896 lunches. This month it served 1,000 lunches. How many more lunches did it serve this month than last?
   (1) 204   (2) 104   (3) 114   (4) 214

7. Exactly 423 men work in a mill. If 85 men are laid off, how many men will still work in the mill?
   (1) 448   (2) 348   (3) 338   (4) 438

8. Last year 26,489 cars came off the assembly line of a motor plant. This year 31,500 cars came off the assembly line. How many more cars came off the assembly line this year?
   (1) 15,121   (2) 5,111   (3) 5,101   (4) 5,011

9. The total taken in at Belduct race track yesterday was $99,000. Today the total was $75,998. How much less was taken in today?
   (1) $23,002   (2) $23,012   (3) $23,102   (4) $23,112

10. In New York 51,246 workers went on strike. In California 31,484 workers went on strike. How many more workers went on strike in New York than in California?
    (1) 29,762   (2) 19,762   (3) 18,862   (4) 20,762

Check your answers on page 85.

7. MULTIPLICATION

There was an old woman from Snow
Who when asked for her age would crow:
"For the answer you seek,
Take the days in the week,
Multiply them by 9; then you'll know."

How old was she? There are several ways to find out. One way is to add. You know there are 7 days in one week. To find out what 9 sevens equal, you could __add__:

```
    7
    7
    7
    7
    7
    7
    7
    7
 +  7
 ─────
   63
```

But multiplying is much quicker:

```
    7
 ×  9
 ─────
   63
```

Multiplication is a quick way to add the same number many times. The sign for multiplication is an ×. It is sometimes called a times sign. Whenever you see this sign, you must multiply. The answer you get when you multiply is called the PRODUCT.

The multiplication table on the next page can help you answer problems. The shaded numbers are the numbers you multiply.

✕	1	2	3	4	5	6	7	8	9
1	1	2	3	4	5	6	7	8	9
2	2	4	6	8	10	12	14	16	18
3	3	6	9	12	15	18	21	24	27
4	4	8	12	16	20	24	28	32	36
5	5	10	15	20	25	30	35	40	45
6	6	12	18	24	30	36	42	48	54
7	7	14	21	28	35	42	49	56	63
8	8	16	24	32	40	48	56	64	72
9	9	18	27	36	45	54	63	72	81

Look at this example:

$$\begin{array}{r} 3 \\ \times\, 2 \end{array}$$

Find 3 in the top shaded row of the chart. Find 2 in the shaded left column. Read down from the 3 and across from the 2. Where they meet is the answer. $3 \times 2 = 6$.

Try It (Answers on page 86.)

Multiply column A. Then multiply column B.

A	B
1. $4 \times 6 =$	1. $6 \times 4 =$
2. $9 \times 8 =$	2. $8 \times 9 =$
3. $5 \times 7 =$	3. $7 \times 5 =$
4. $6 \times 3 =$	4. $3 \times 6 =$
5. $2 \times 7 =$	5. $7 \times 2 =$

Did you notice that column A and column B have the same answers? The order of the numbers you multiplied was changed from column A to column B. If you can work column A, you can also work column B. In multiplication, it does not matter which number comes first.

The chart on page 40 is helpful when you are learning to multiply. But you cannot have the chart in front of you everywhere you go. With practice, you will not have to use the chart as often. Eventually, you will learn the whole chart.

MULTIPLYING BY ZERO

Any number multiplied by zero equals zero. Look at the example below:

$$0 + 0 + 0 + 0 + 0 = 0$$
$$\text{or, } 5 \times 0 = 0$$

Try It (Answers on page 86.)

1.	5	2.	6	3.	0	4.	8	5.	0
	× 0		× 0		× 3		× 0		× 2

MULTIPLYING BY ONE

Any number multiplied by one equals that number. Look at the example below.

$$1 + 1 + 1 + 1 + 1 + 1 = 6$$
$$\text{or } 6 \times 1 = 6$$

Try It (Answers on page 86.)

1.	9	2.	7	3.	3	4.	1	5.	6
	× 1		× 1		× 1		× 8		× 1

MULTIPLYING TENS NUMBERS

To multiply a number that has tens and ones places, multiply each column separately. Begin by multiplying the ones place. Then multiply the tens place. Look at the example below:

		Step 1		Step 2	
		Tens	Ones	Tens	Ones
24		2	4	2	4
× 2	×		2	×	2
			8	4	8

Step 1—Multiply the ones number: $2 \times 4 = 8$. Write your answer in the ones column.

Step 2—Multiply the tens number: $2 \times 2 = 4$. Write your answer in the tens column.

Try It (Answers on page 86.)

1.	13	2.	21	3.	12	4.	30	5.	11
	× 3		× 2		× 2		× 2		× 7

6.	23	7.	41	8.	22	9.	98	10.	14
	× 3		× 2		× 4		× 1		× 2

11.	20	12.	31	13.	43	14.	12	15.	22
	× 4		× 3		× 2		× 3		× 3

CARRYING TO THE TENS COLUMN

You might have a multiplication answer in the ones column that is ten or more. Carry the tens number in the answer. Place it in the tens column. The carried number is always added after the tens number is multiplied. Look at the example below:

	Step 1	Step 2
3 6 × 2	¹ 3 6 × 2 — 2	¹ 3 6 × 2 — 7 2

Step 1—Multiply the ones number: $2 \times 6 = 12$. Carry the 10 ones as 1 ten and place it in the tens column. Leave the 2 ones in the ones column.

Step 2—Multiply the tens number: $2 \times 3 = 6$. Add the number of tens you carried: $6 + 1 = 7$.

Try It (Answers on page 86.)

1.	24	2.	13	3.	37	4.	46	5.	14
	× 3		× 5		× 2		× 2		× 6

6. 16	7. 28	8. 25	9. 39	10. 17
× 4	× 3	× 3	× 2	× 4

11. 18	12. 26	13. 37	14. 19	15. 18
× 3	× 3	× 2	× 4	× 4

MULTIPLYING HUNDREDS NUMBERS

When a number has ones, tens, and hundreds columns, multiply each column separately. Look at the example below:

	Step 1	Step 2	Step 3

	312		H	T	O		H	T	O		H	T	O
	× 2	×	3	1	2	×	3	1	2	×	3	1	2
					2				2				2
					4			2	4		6	2	4

Step 1—Multiply the number in the ones column: $2 \times 2 = 4$. Write your answer in the ones column.

Step 2—Multiply the number in the tens column: $2 \times 1 = 2$. Write your answer in the tens column.

Step 3—Multiply the number in the hundreds column: $2 \times 3 = 6$. Write your answer in the hundreds column.

Try It (Answers on page 86.)

1. 313	2. 231	3. 143	4. 202	5. 423
× 3	× 3	× 2	× 3	× 2

6. 321	7. 230	8. 221	9. 333	10. 204
× 2	× 3	× 3	× 2	× 2

CARRYING TO THE HUNDREDS COLUMN

Numbers may be carried and placed in the hundreds column, as they are carried and placed in the tens column. Look at this example:

	Step 1	Step 2	Step 3
346 × 4	346 × 4 4	$\overset{2}{3}46$ × 4 84	$\overset{1}{3}46$ × 4 1,384

Step 1—Multiply: $4 \times 6 = 24$. The 2 tens are carried and placed in the tens column.

Step 2—Multiply: $4 \times 4 = 16$. Add the 2 tens you carried: $2 + 16 = 18$. The 1 in this answer is 10 tens or 1 hundred. Thus, the 1 is carried and placed in the hundreds column.

Step 3—Multiply: $4 \times 3 = 12$. Add the 1 you carried: $12 + 1 = 13$.

Try It (Answers on page 86.)

1. 167 × 4	2. 143 × 6	3. 198 × 5	4. 474 × 2	5. 185 × 5
6. 307 × 3	7. 242 × 4	8. 271 × 3	9. 139 × 2	10. 370 × 2
11. 364 × 5	12. 781 × 7	13. 639 × 8	14. 977 × 6	15. 408 × 9

MULTIPLYING THOUSANDS NUMBERS

When a number has columns larger than the hundreds column, you still multiply each column separately. Look at the example below:

	Step 1	Step 2	Step 3	Step 4

2,134 × 2	Th	H	T	O		Th	H	T	O		Th	H	T	O		Th	H	T	O
	2	1	3	4		2	1	3	4		2	1	3	4		2	1	3	4
×				2	×				2	×				2	×				2
				8				6	8			2	6	8		4	2	6	8

Step 1—Multiply the ones column: $2 \times 4 = 8$.

Step 2—Multiply the tens column: $2 \times 3 = 6$.

Step 3—Multiply the hundreds column: $2 \times 1 = 2$.

Step 4—Multiply the thousands column: $2 \times 2 = 4$.

Try It (Answers on page 86.)

1. 3,021	2. 1,324	3. 3,032	4. 1,220	5. 1,102
$\times$ 2	$\times$ 2	$\times$ 3	$\times$ 4	$\times$ 3

6. 2,103	7. 12,020	8. 3,413	9. 2,331	10. 32,230
$\times$ 2	$\times$ 4	$\times$ 2	$\times$ 2	$\times$ 3

CARRYING TO THE THOUSANDS COLUMN

Numbers are carried and placed in columns larger than the hundreds column in the same way they are carried and placed in the hundreds column.

	Step 1	Step 2	Step 3	Step 4
5,675 $\times$ 3	$5,6\overset{1}{7}5$ $\times$ 3 5	$5,\overset{2}{6}\overset{1}{7}5$ $\times$ 3 25	$\overset{2}{5},\overset{2}{6}\overset{1}{7}5$ $\times$ 3 025	$\overset{2}{5},\overset{2}{6}\overset{1}{7}5$ $\times$ 3 17,025

Step 1—Multiply: $3 \times 5 = 15$. Carry the 1 and place it in the tens column.

Step 2—Multiply: $3 \times 7 = 21$. Add: $21 + 1 = 22$. Carry the 2 and place it in the hundreds column.

Step 3—Multiply: $3 \times 6 = 18$. Add: $18 + 2 = 20$. Carry the 2 and place it in the thousands column.

Step 4—Multiply: $3 \times 5 = 15$. Add: $15 + 2 = 17$.

Try It (Answers on page 87.)

1. 1,347	2. 2,687	3. 3,906	4. 2,851	5. 2,609
$\times$ 4	$\times$ 3	$\times$ 2	$\times$ 3	$\times$ 3

6. 2,486
    ×   4

7. 24,579
    ×   2

8. 12,948
    ×   5

9. 3,570
    ×   2

10. 2,717
    ×   3

11. 4,234
    ×   8

12. 7,810
    ×   9

13. 5,609
    ×   5

14. 3,042
    ×   6

15. 2,996
    ×   7

WORKING WITH WORD PROBLEMS

Read the Steps for Solving Word Problems on page xii again. Then work these problems. Find the correct answer.

Here are some important clue words. They can help you solve word problems in multiplication. These clue words usually mean you must multiply.

1. **How much** (or **How many**) **for a larger quantity**—The question might ask: "If candy costs $2 for one box, how much will 6 boxes cost?"

2. **Times**—The question might say, "Five <u>times</u> as many people work in the factory."

3. **At**—The question might ask, "What is the cost of 3 dozen eggs <u>at</u> 90 cents a dozen?"

Now, work these word problems.

1. A policeman gave out nine parking tickets a day for six days. How many parking tickets did he give altogether?
    (1) 72   (2) 48   (3) 54   (4) 63

2. Oscar Robertson scored 28 points in each of six basketball games. What is the total number of points he scored?
    (1) 128   (2) 186   (3) 166   (4) 168

3. Roberta's new car goes 17 miles on one gallon of gas. How many miles can she drive on nine gallons?
    (1) 153   (2) 144   (3) 93   (4) 162

4. The Snyders traveled 350 miles a day for seven days. How many miles did they travel?
    (1) 2,150 (2) 2,450 (3) 2,400 (4) 2,457

5. 1,487 people cross the border between Mexico and the United States each day. How many people cross the border in eight days?
   (1) 8,296   (2) 11,896   (3) 11,296   (4) 11,246

6. Four cities have Head Start programs. Each program has eight classes. There are 18 children in each class. How many children are in the four programs?
   (1) 72   (2) 144   (3) 576   (4) 32

7. Capitol Movies has 559 seats. It has two shows a day. Every seat is filled at each show. How many people go to the movies in six days?
   (1) 6,708   (2) 3,354   (3) 3,304   (4) 1,118

8. Midtown Hospital serves three meals a day. There are 456 patients in the hospital. How many meals are served in one week?
   (1) 1,368   (2) 3,192   (3) 9,576   (4) 4,560

9. Rick works as a delivery man. He earns $6 an hour. This week he worked 45 hours. How much did he make?
   (1) $24   (2) $240   (3) $270   (4) $27

10. There are 12 eggs to a dozen. Sam's Lunch Spot ordered 12 dozen eggs. How many eggs did they order?
    (1) 12   (2) 24   (3) 144   (4) 36

Check your answers on page 87.

8. MORE ON MULTIPLICATION

MULTIPLYING BY TENS NUMBERS

You know that multiplication is repeated addition. But addition is very slow. How much are twenty-three 31's? You could add 31 twenty-three times. An easier way is to multiply 31 by 23. Look at the example below:

		Step 1	Step 2	Step 3

	Step 1	Step 2	Step 3
$\begin{array}{r} 31 \\ \times\ 23 \end{array}$	H T O 3 1 × 2 3 9 3	H T O 3 1 × 2 3 9 3 6 2 0	H T O 3 1 × 2 3 9 3 + 6 2 0 7 1 3

Step 1—Multiply: $3 \times 31 = 93$.

Step 2—Multiply: $2 \times 31 = 62$. Since you are multiplying by a <u>tens</u> number, <u>start your answer in the tens column.</u> You can use a zero in the ones column as a place holder.

Step 3—To find your final answer, <u>add</u>: $93 + 620 = 713$.

Look at this example. You must carry in steps 1 and 2.

	Step 1	Step 2	Step 3
$\begin{array}{r} 634 \\ \times\ 67 \end{array}$	Th H T O $\overset{2}{6}\ \overset{2}{3}\ 4$ × 6 7 4 4 3 8	TTh Th H T O $\overset{2}{6}\ \overset{2}{3}\ 4$ × 6 7 4 4 3 8 0 3 8 0 4 0	TTh Th H T O 6 3 4 × 6 7 4 4 3 8 0 3 8 0 4 0 4 2 4 7 8

Step 1—Multiply: $7 \times 4 = 28$. Carry the 2. Multiply: $7 \times 3 = 21$. Add the 2 you carried: $21 + 2 = 23$. Carry the 2. Multiply: $7 \times 6 = 42$. Add the 2 you carried: $42 + 2 = 44$.

Step 2—Multiply: $6 \times 4 = 24$. Carry the 2. Start your answer in the tens column. You can use a zero as a place holder. Multiply: $6 \times 3 = 18$. Add the 2 you carried: $18 + 2 = 20$. Carry the 2. Multiply: $6 \times 6 = 36$. Add the 2 you carried: $36 + 2 = 38$.

Step 3—Add to find the final answer:

$$\begin{array}{r} 4438 \\ + 38040 \\ \hline 42478 \end{array}$$

Try It (Answers on page 87.)

1. $\begin{array}{r} 42 \\ \times 12 \\ \hline \end{array}$
2. $\begin{array}{r} 33 \\ \times 23 \\ \hline \end{array}$
3. $\begin{array}{r} 21 \\ \times 20 \\ \hline \end{array}$
4. $\begin{array}{r} 40 \\ \times 22 \\ \hline \end{array}$
5. $\begin{array}{r} 32 \\ \times 11 \\ \hline \end{array}$

6. $\begin{array}{r} 68 \\ \times 34 \\ \hline \end{array}$
7. $\begin{array}{r} 27 \\ \times 45 \\ \hline \end{array}$
8. $\begin{array}{r} 97 \\ \times 25 \\ \hline \end{array}$
9. $\begin{array}{r} 627 \\ \times 41 \\ \hline \end{array}$
10. $\begin{array}{r} 978 \\ \times 68 \\ \hline \end{array}$

11. $\begin{array}{r} 4,725 \\ \times 59 \\ \hline \end{array}$
12. $\begin{array}{r} 803 \\ \times 58 \\ \hline \end{array}$
13. $\begin{array}{r} 275 \\ \times 84 \\ \hline \end{array}$
14. $\begin{array}{r} 2,963 \\ \times 54 \\ \hline \end{array}$
15. $\begin{array}{r} 6,897 \\ \times 98 \\ \hline \end{array}$

16. $\begin{array}{r} 397 \\ \times 76 \\ \hline \end{array}$
17. $\begin{array}{r} 3,563 \\ \times 63 \\ \hline \end{array}$
18. $\begin{array}{r} 9,089 \\ \times 35 \\ \hline \end{array}$
19. $\begin{array}{r} 781 \\ \times 26 \\ \hline \end{array}$
20. $\begin{array}{r} 1,322 \\ \times 57 \\ \hline \end{array}$

21. $\begin{array}{r} 20,703 \\ \times 49 \\ \hline \end{array}$
22. $\begin{array}{r} 5,280 \\ \times 72 \\ \hline \end{array}$
23. $\begin{array}{r} 1,979 \\ \times 33 \\ \hline \end{array}$
24. $\begin{array}{r} 6,428 \\ \times 53 \\ \hline \end{array}$
25. $\begin{array}{r} 4,300 \\ \times 68 \\ \hline \end{array}$

MULTIPLYING BY HUNDREDS NUMBERS

To multiply by hundreds numbers, multiply each column separately. When you multiply by the hundreds column number, place the first number in your answer in the hundreds column. You can use two zeros as place holders. Look at the example below:

	Step 1	Step 2	Steps 3-4

	H	T	O
	$\overset{1}{2}$	4	1
×	1	6	3
	7	2	3

241
× 163

Step 1 table:

	H	T	O
	$\overset{1}{2}$	4	1
×	1	6	3
	7	2	3

Step 2 table:

TTh	Th	H	T	O
		$\overset{2}{2}$	4	1
×		1	6	3
		7	2	3
1	4	4	6	0

Steps 3-4 table:

TTh	Th	H	T	O
		2	4	1
×		1	6	3
		7	2	3
1	4	4	6	0
2	4	1	0	0
3	9	2	8	3

Step 1—Multiply: $3 \times 1 = 3$. Multiply: $3 \times 4 = 12$. Carry the 1. Multiply $3 \times 2 = 6$. Add the 1 you carried: $6 + 1 = 7$.

Step 2—Multiply: $6 \times 1 = 6$. Since you are multiplying by a tens number, start your answer in the tens column. You can use a zero in the ones column as a place holder. Multiply: $6 \times 4 = 24$. Carry the 2. Multiply $6 \times 2 = 12$. Add the 2 you carried: $12 + 2 = 14$.

Step 3—Multiply $1 \times 1 = 1$. Since you are multiplying by a hundreds number, start your answer in the hundreds column. You can use two zeros as place holders. Multiply $1 \times 4 = 4$. Multiply: $1 \times 2 = 2$.

Step 4—Add to find your final answer:

$$
\begin{array}{r}
723 \\
14,460 \\
+ 24,100 \\
\hline
39,283
\end{array}
$$

Try It (Answers on page 87.)

1. 314   2. 221   3. 142   4. 323   5. 340
  × 212   × 314   × 121   × 321   × 234

6. 678	7. 726	8. 465	9. 937	10. 358
× 448	× 382	× 571	× 571	× 269

11. 3,426	12. 5,963	13. 7,809	14. 4,463	15. 8,093
× 852	× 746	× 682	× 987	× 527

16. 5,280	17. 6,333	18. 10,298	19. 3,005	20. 1,213
× 365	× 182	× 615	× 970	× 456

21. 5,423	22. 2,000	23. 7,389	24. 3,400	25. 4,682
× 235	× 481	× 267	× 836	× 453

WORKING WITH WORD PROBLEMS

Read the Steps for Solving Word Problems on page xii again. Then work these problems. Find the correct answer.

1. 56 people are working in the tomato field. Each person picks 25 bushels of tomatoes a day. How many bushels of tomatoes will be picked today?
   (1) 130   (2) 140   (3) 392   (4) 1,400

2. Mr. Warwick collects $45 a week unemployment compensation. He can collect this for 26 weeks. How much can he collect?
   (1) $340   (2) $360   (3) $1,150 (4) $1,170

3. Sharon pays $215 a month rent. How much will she pay in rent in 2 years? (1 year = 12 months)
   (1) $5,160 (2) $5,040 (3) $2,580 (4) $1,290

4. Earl and his wife earn a total of $290 a week. How much do they earn together in a year? (1 year = 52 weeks)
   (1) $14,880 (2) $15,132 (3) $15,080 (4) $20,300

5. 35 people are working on a voter registration drive. Each person hopes to register at least 250 people. How many people do they hope to register?
   (1) 875   (2) 8,750   (3) 2,000   (4) 87,500

6. A labor union has 4,987 members. Each member pays dues of $48 a year. How much in dues are collected each year by the union?
   (1) $59,844   (2) $60,844   (3) $239,376   (4) $240,376

7. There are 60 minutes in an hour. There are 24 hours in a day. There are 365 days in a year. How many minutes are there in a year?
   (1) 8,760   (2) 20,160   (3) 21,900   (4) 525,600

8. Joanne bought a used car for $3,000. She agreed to make monthly payments of $145 for two years. How much will she pay for the car?
   (1) $3,000   (2) $3,470   (3) $3,480   (4) $3,490

9. Carol and Richard bought a dining room set for $395. They agreed to make 36 monthly payments of $13 a month. This includes the service charge for buying "on time." How much more will they pay by buying "on time"?
   (1) $13   (2) $73   (3) $82   (4) $108

10. Riverton sheet metal factory has 5 work crews. Each work crew has 25 members. Each person cuts 2,394 sheets of metal a week. What is the total number of sheets cut in a week?
    (1) 11,970 (2) 43,390 (3) 59,850 (4) 299,250

Check your answers on page 88.

9. DIVISION

"How will we divide the loot?" asked the first thief.

"Well," said his friend, "we've got $100, and there are two of us. If we divide the $100 by 2, we will each have $50."

The first thief thought about it. Then he took the money and ran. He didn't have to divide anything. But you cannot run away from division. You use division to solve problems every day.

One sign for division looks like a dash between two dots. Here is an example of a division problem:

$$6 \div 2 =$$

This is read "6 divided by 2." The number <u>after</u> this division sign is always divided <u>into</u> the number <u>before</u> the sign.

Another sign for division looks like this:

$$2\overline{)6}$$

This is also read, "6 divided by 2." The number <u>outside</u> this sign is always divided <u>into</u> the number <u>inside</u> the sign.

In division, the answer is called the QUOTIENT.

You can divide if you know how to multiply. For example: $6 \div 2 = 3$ since $2 \times 3 = 6$. Look at these problems:

Problem:	Think:	Answer:
$3\overline{)6}$	$3 \times 2 = 6$	$3\overline{)6}^{2}$
$2\overline{)18}$	$2 \times 9 = 18$	$2\overline{)18}^{9}$
$5\overline{)20}$	$5 \times 4 = 20$	$5\overline{)20}^{4}$

Try It (Answers on page 88.)

Problem:	Think:	Problem:	Think:
1. $2\overline{)8}$	$2 \times \underline{\quad} = 8$	2. $5\overline{)35}$	$5 \times \underline{\quad} = 35$

53

3. $4\overline{)12}$ $4 \times \underline{} = 12$ 4. $7\overline{)56}$ $7 \times \underline{} = 56$

5. $9\overline{)45}$ $9 \times \underline{} = 45$ 6. $8\overline{)48}$ $8 \times \underline{} = 48$

7. $3\overline{)27}$ $3 \times \underline{} = 27$ 8. $4\overline{)16}$ $4 \times \underline{} = 16$

9. $4\overline{)24}$ $4 \times \underline{} = 24$ 10. $5\overline{)25}$ $5 \times \underline{} = 25$

11. $8\overline{)72}$ $8 \times \underline{} = 72$ 12. $3\overline{)15}$ $3 \times \underline{} = 15$

13. $9\overline{)45}$ $9 \times \underline{} = 45$ 14. $7\overline{)49}$ $7 \times \underline{} = 49$

15. $6\overline{)18}$ $6 \times \underline{} = 18$ 16. $4\overline{)28}$ $4 \times \underline{} = 28$

17. $5\overline{)30}$ $5 \times \underline{} = 30$ 18. $2\overline{)16}$ $2 \times \underline{} = 16$

19. $9\overline{)63}$ $9 \times \underline{} = 63$ 20. $8\overline{)64}$ $8 \times \underline{} = 64$

DIVIDING INTO TENS NUMBERS

To divide into tens numbers, divide from the left. First divide into the tens number. Then divide into the ones number. Look at this example:

	Step 1	Step 2
$2\overline{)86}$	$\overset{4}{2\overline{)86}}$	$\overset{43}{2\overline{)86}}$

Step 1— Divide into the tens number: $8 \div 2 = 4$. Write the 4 over the 8.

Step 2— Divide into the ones number: $6 \div 2 = 3$. Write the 3 over the 6.

CHECKING DIVISION ANSWERS

Check your division answers by multiplying. Multiply the number you divided by and your answer. Look at the example below:

Problem: Check:

$$2\overline{)86} \atop 43$$

$$\begin{array}{r} 43 \\ \times\ 2 \\ \hline 86 \end{array}$$

Try It (Answers on page 88.)

Divide. Check your answers.

1. $4\overline{)48}$ 2. $2\overline{)84}$ 3. $2\overline{)48}$ 4. $3\overline{)63}$ 5. $2\overline{)26}$

6. $4\overline{)84}$ 7. $2\overline{)62}$ 8. $3\overline{)96}$ 9. $2\overline{)22}$ 10. $6\overline{)66}$

ZERO IS A PLACE HOLDER

Zero at the end of a number is a place holder. You know that 60 is not the same as 6. When you divide into a number that ends in zero, don't forget to put the zero in your answer. Look at this example:

$$3\overline{)30} \atop 10$$

$$\begin{array}{r} 10 \\ \times\ 3 \\ \hline 30 \end{array}$$

Try It (Answers on page 89.)

Divide. Check your answers.

1. $2\overline{)20}$ 2. $2\overline{)40}$ 3. $3\overline{)60}$ 4. $7\overline{)70}$ 5. $4\overline{)40}$

6. $2\overline{)60}$ 7. $4\overline{)80}$ 8. $5\overline{)50}$ 9. $2\overline{)80}$ 10. $3\overline{)90}$

DIVIDING INTO HUNDREDS NUMBERS

To divide into hundreds place numbers, divide from the left. First divide into the hundreds number. Then divide into the tens number. Finally, divide into the ones number. Look at the example below:

	Step 1	Step 2	Step 3
4)884	2 4)884	2 2 4)884	2 2 1 4)884

Step 1—Divide into the hundreds number: $8 \div 4 = 2$. Write the 2 over the hundreds 8.

Step 2—Divide into the tens number: $8 \div 4 = 2$. Write the 2 over the tens 8.

Step 3—Divide into the ones number: $4 \div 4 = 1$. Write the 1 over the 4.

Check your answer by multiplying:

Problem: Check:

$$\begin{array}{r} 221 \\ 4\overline{)884} \end{array} \qquad \begin{array}{r} 221 \\ \times \quad 4 \\ \hline 884 \end{array}$$

Try It (Answers on page 89.)

Divide. Check your answers.

1. 3)633   2. 4)488   3. 2)468   4. 2)242

5. 4)848   6. 3)996   7. 5)555   8. 2)862

9. 3)336   10. 2)422   11. 3)939   12. 2)684

ZERO IN THE MIDDLE

You know that zero is a place holder. You also know that any number divided into zero equals zero. If there is a zero in your answer, you must put it in as a place holder. Look at this example:

	Step 1	Step 2	Step 3
3)306	1 3)306	10 3)306	102 3)306

Step 1—Divide: $3 \div 3 = 1$.

Step 2—Divide: $0 \div 3 = 0$.
　　　　Remember to put the zero in your answer.

Step 3—Divide: $6 \div 3 = 2$.

Check your answer by multiplying:

Problem:　　　　Check:

$$
\begin{array}{r}
102 \\
3)\overline{306}
\end{array}
\qquad
\begin{array}{r}
102 \\
\times\ \ \ 3 \\
\hline
306
\end{array}
$$

Try It (Answers on page 89.)

Divide. Check your answers.

1. 4)804　　　2. 2)206　　　3. 3)609　　　4. 6)606　　　5. 3)903

6. 2)602　　　7. 3)306　　　8. 4)804　　　9. 5)505　　　10. 2)406

DIVIDING INTO THOUSANDS NUMBERS

To divide into thousands place numbers, start at the left. First, divide into the thousands. Then divide into the hundreds, tens, and ones numbers. Look at the example below:

	Step 1	Step 2	Step 3	Step 4
2)6,842	3 2)6,842	3 4 2)6,842	3 42 2)6,842	3,421 2)6,842

Step 1—Divide into the thousands number: $6 \div 2 = 3$.

Step 2—Divide into the hundreds number: $8 \div 2 = 4$.

Step 3—Divide into the tens number: $4 \div 2 = 2$.

Step 4—Divide into the ones number: $2 \div 2 = 1$.

Check your answer by multiplying:

$$
\begin{array}{cc}
\text{Problem:} & \text{Check:} \\
3,421 & 3,421 \\
2\overline{)6,842} & \times\quad 2 \\
& \overline{6,842}
\end{array}
$$

Try It (Answers on page 89.)

Divide. Check your answers.

1. $3\overline{)9,369}$ 2. $4\overline{)4,808}$ 3. $2\overline{)6,428}$

4. $6\overline{)6,606}$ 5. $3\overline{)3,993}$ 6. $2\overline{)4,242}$

7. $3\overline{)6,639}$ 8. $7\overline{)7,777}$ 9. $4\overline{)8,448}$

WHEN IT LOOKS AS IF YOU CANNOT DIVIDE

Sometimes you cannot divide into the first number of a problem. The first number may be too small. You must divide into the first <u>two</u> numbers. Look at the example below:

	Step 1	Step 2	Step 3
$3\overline{)123}$	$3\overline{)123}$	$\begin{array}{r}4\\3\overline{)123}\end{array}$	$\begin{array}{r}41\\3\overline{)123}\end{array}$

Step 1—3 cannot divide into 1.

Step 2—3 can divide into 12: $12 \div 3 = 4$. Write the 4 over the 2.

Step 3—3 can divide into 3: $3 \div 3 = 1$. Write the 1 over the 3.

Check your answer by multiplying:

$$
\begin{array}{cc}
\text{Problem:} & \text{Check:} \\
41 & 41 \\
3\overline{)123} & \times\ 3 \\
& \overline{123}
\end{array}
$$

Try It (Answers on page 89.)

Divide. Check your answers.

1. 4)168 2. 5)255 3. 7)567 4. 3)246 5. 6)426

6. 8)728 7. 4)284 8. 3)219 9. 7)427 10. 2)108

Sometimes the second or third number is too small to divide. A zero holds the place when a number is too small to be divided. Look at the example below:

	Step 1	Step 2	Step 3
3)612	2 3)612	20 3)612	204 3)612

Step 1—3 can divide into 6: 6 ÷ 3 = 2. Write the 2 over the 6.

Step 2—3 cannot divide into 1. Write a 0 over the 1.

Step 3—3 can divide into 12: 12 ÷ 3 = 4. Write the 4 over the 2.

Check your answer by multiplying:

Problem: Check:
    204   204
3)612   ×   3
                   612

Try It (Answers on page 89.)

Divide. Check your answers.

1. 4)816 2. 5)515 3. 6)618 4. 8)856 5. 3)918

6. 7)721 7. 2)814 8. 9)954 9. 3)624 10. 9)945

NUMBERS THAT CANNOT BE DIVIDED EVENLY

Can you divide 5 pencils evenly between 2 people? How many pencils does each person get? How many pencils are left over? Look at the example below:

	Step 1	Step 2
$2\overline{)5}$	$2\overline{)5}$ over: 2	$\begin{array}{r} 2\ r\ 1 \\ 2\overline{)\ 5} \\ -\ 4 \\ \hline 1 \end{array}$

Step 1—2 does not divide into 5 evenly. 2 does divide into 4 evenly: $4 \div 2 = 2$. Write the 2 over the 5.

Step 2—There is 1 left over: $5 - 4 = 1$. 1 is the remainder. Write <u>r1</u> after the 2.

Each person gets 2 pencils. There is 1 pencil left over. The part left is called the <u>remainder</u>. The letter <u>r</u> is a short way of writing remainder.

Check your answer by <u>multiplying</u> and then <u>adding</u> the remainder.

Problem: Check:

$\begin{array}{r} 2\ r\ 1 \\ 2\overline{)5} \\ 4 \\ \hline 1 \end{array}$
$\begin{array}{r} 2 \\ \times\ 2 \\ \hline 4 \end{array}$ Add the remainder. $\begin{array}{r} 4 \\ +\ 1 \\ \hline 5 \end{array}$

Try It (Answers on page 89.)

Divide. Each one will have a remainder. Check your answers.

1. $3\overline{)7}$ 2. $4\overline{)9}$ 3. $2\overline{)7}$ 4. $4\overline{)6}$ 5. $5\overline{)11}$

6. $6\overline{)14}$ 7. $7\overline{)23}$ 8. $8\overline{)42}$ 9. $5\overline{)38}$ 10. $4\overline{)27}$

11. $9\overline{)65}$ 12. $6\overline{)58}$ 13. $3\overline{)25}$ 14. $8\overline{)68}$ 15. $5\overline{)49}$

16. $7\overline{)50}$ 17. $9\overline{)75}$ 18. $5\overline{)24}$ 19. $4\overline{)26}$ 20. $8\overline{)50}$

LONG DIVISION—REMAINDERS IN THE TENS PLACE

Sometimes the tens number cannot be divided evenly. Look at this example:

	Step 1	Step 2	Step 3	Step 4	Step 5
$2\overline{)34}$	$\dfrac{1}{2\overline{)34}}$	$2\overline{)\begin{array}{r}1\\34\\2\end{array}}$	$2\overline{)\begin{array}{r}1\\34\\2\\\hline1\end{array}}$	$2\overline{)\begin{array}{r}1\\34\\2\\\hline14\end{array}}$	$2\overline{)\begin{array}{r}17\\34\\2\\\hline14\\14\\\hline0\end{array}}$

Step 1—Divide the tens number: $3 \div 2 = 1$. Write the 1 above the 3.

Step 2—Multiply: $1 \times 2 = 2$. Write the 2 under the 3.

Step 3—Subtract: $3 - 2 = 1$. Write 1 under the 2.

Step 4—Bring down the next number. Write the 4 next to the remainder 1.

Step 5—Divide: $14 \div 2 = 7$. Write the 7 above the 4. Multiply: $7 \times 2 = 14$. Write 14 under the 14 remainder. Subtract: $14 - 14 = 0$. There is nothing left to divide. The problem is finished.

Check your answer by multiplying:

Problem: Check:

$$2\overline{)\begin{array}{r}17\\34\end{array}} \qquad \begin{array}{r}17\\ \times\ 2\\\hline34\end{array}$$

Try It (Answers on page 89.)

1. $3\overline{)48}$ 2. $6\overline{)78}$ 3. $8\overline{)96}$ 4. $4\overline{)64}$ 5. $5\overline{)75}$

6. $7\overline{)84}$ 7. $4\overline{)56}$ 8. $3\overline{)55}$ 9. $5\overline{)85}$ 10. $7\overline{)91}$

11. $6\overline{)88}$ 12. $5\overline{)65}$ 13. $4\overline{)96}$ 14. $3\overline{)58}$ 15. $4\overline{)72}$

DIVIDING LARGER NUMBERS

You need to use long division on some large numbers. The steps are the same. Look at this example:

	Step 1	Step 2	Step 3	Step 4
6)978	1 6)978	1 6)978 6	1 6)978 6 3	1 6)978 6 37

Step 1—Divide: 9 ÷ 6 = 1. Write the 1 above the 9.

Step 2—Multiply: 1 × 6 = 6. Write the 6 under the 9.

Step 3—Subtract: 9 − 6 = 3. Write the 3 under the 6.

Step 4—Bring down the next number: 7. Write the 7 next to the remainder 3.

Repeat steps 1 - 4 again

Divide: 37 ÷ 6 = 6
Multiply: 6 × 6 = 36
Subtract: 37 − 36 = 1

```
      16
6)978
   6
   37
   36
    1
```

Bring down the next number. Repeat the steps again.

Divide: 18 ÷ 6 = 3
Multiply: 3 × 6 = 18
Subtract: 18 − 18 = 0

```
      163
6)978
   6
   37
   36
    18
    18
     0
```

There are no more numbers to bring down and the remainder is 0. The problem is finished.

The problem works out evenly. There is no remainder. Check your answer by multiplying.

Problem:	Check:
163	163
6)978	× 6
	978

Try It (Answers on page 90.)

1. 3)723   2. 2)554   3. 4)972   4. 5)620   5. 6)728

6. 2)998   7. 8)912   8. 3)775   9. 6)373   10. 2)712

11. 3)2,922   12. 8)7,536   13. 6)2,598   14. 5)4,280   15. 2)7,112

16. 7)5,971   17. 4)5,104   18. 9)8,487   19. 7)4,454   20. 3)5,421

WORKING WITH WORD PROBLEMS

Read the Steps for Solving Word Problems on page xii again. Then work these problems. Find the correct answer.

Here are some important clue words. They can help you solve division word problems.

1. **Divide**—The question might say, "The money was divided equally among four people."

2. **Each**—The question might ask, "How much does each one cost?"

Now, work these word problems.

1. Four people work in the sales office. They have eight doughnuts for coffee break today. How many doughnuts can each person have?
   (1) 1   (2) 2   (3) 3   (4) 4

2. Maria wants to buy grapefruit juice. It is on sale at three cans for 99¢. How much does each can cost?
   (1) 30¢   (2) 31¢   (3) 32¢   (4) 33¢

3. There are 202 workers in the Acme factory. There are two shifts. The same number work on each shift. How many people work on each shift?
   (1) 100   (2) 101   (3) 202   (4) 404

4. There are 20 cigarettes in a pack. Mr. Pollack smokes two cigarettes an hour. How many hours will it take him to smoke a pack?
   (1) 9   (2) 10   (3) 20   (4) 40

5. There are 30 slices in a loaf of Wonderlite bread. It takes two slices of bread for a sandwich. How many sandwiches can you make from a loaf?
   (1) 10   (2) 15   (3) 25   (4) 30

6. Ed has eight payments left on his car. He has 240 dollars to pay. How much is each payment?
   (1) $8   (2) $20   (3) $24   (4) $30

7. Tickets to the Saturday soccer game were $3 each. The ticket sellers collected $3,246. How many tickets were sold?
   (1) 182   (2) 1,082   (3) 1,182   (4) 982

8. Kim found a TV set on sale. It cost $200, including interest. She put down $40. She signed a contract to make eight equal monthly payments. How much will she pay each month?
   (1) $20   (2) $25   (3) $35   (4) $40

9. There are 1,561 people working in the seven buildings of the Seaway factory. The same number work in each building. How many people work in each building?
   (1) 123   (2) 333   (3) 223   (4) 233

10. The prize money for the state lottery was $63,027. It was divided equally among nine winners. How much did each person get?
    (1) $703   (2) $7,003   (3) $6,003   (4) $73

Check your answers on page 90.
   your answers on page 90.

10. MORE ON DIVISION

DIVIDING BY TENS NUMBERS

How good are you at guessing? Sometimes you must guess to work a division problem. But sometimes guesses are wrong. You must check. Look at this example:

	Step 1	Step 2
$12 \overline{)36}$	$\begin{array}{r} 3 \\ 12 \overline{)36} \end{array}$	$\begin{array}{r} 12 \\ \times\ 3 \\ \hline 36 \end{array}$

Step 1—Guess that $36 \div 12 = 3$. Why guess 3? Because 1, the first number of 12, divides into the first number of 36 (3) three times.

Step 2—Check by multiplying: $3 \times 12 = 36$. There is no remainder: $36 - 36 = 0$. 3 is correct.

Here is an example of a wrong guess:

	Step 1	Step 2
$14 \overline{)42}$	$\begin{array}{r} 4 \\ 14 \overline{)42} \end{array} \quad \begin{array}{r} 14 \\ \times\ 4 \\ \hline 56 \end{array}$	$\begin{array}{r} 3 \\ 14 \overline{)42} \end{array} \quad \begin{array}{r} 14 \\ \times\ 3 \\ \hline 42 \end{array}$
	(does not check)	(checks)

Step 1—Since the first number of 14 (1) divides into the first number of 42 (4) four times, guess 4.

Step 2—Check by multiplying: $4 \times 14 = 56$. 56 is too large. Try the next smaller number, 3. Multiply: $3 \times 14 = 42$. 3 works.

Try It (Answers on page 90.)

Divide. Check your answers.

1. $10 \overline{)60}$ 2. $14 \overline{)28}$ 3. $11 \overline{)33}$ 4. $13 \overline{)39}$ 5. $12 \overline{)48}$

6. 16$\overline{)48}$ 7. 18$\overline{)54}$ 8. 28$\overline{)84}$ 9. 23$\overline{)69}$ 10. 17$\overline{)68}$

11. 33$\overline{)99}$ 12. 15$\overline{)75}$ 13. 13$\overline{)78}$ 14. 19$\overline{)38}$ 15. 21$\overline{)63}$

DIVIDING BY TENS WITH REMAINDERS

It takes two steps to check division problems with remainders. Look at this example:

	Step 1	Step 2	Step 3	Step 4
12$\overline{)38}$	$\begin{array}{r}3\\12\overline{)38}\end{array}$	$\begin{array}{r}3\\12\overline{)38}\\36\end{array}$	$\begin{array}{r}3\\12\overline{)38}\\36\\\hline2\end{array}$	$\begin{array}{r}3\ r\ 2\\12\overline{)38}\\36\\\hline2\end{array}$

Step 1—The first number of 12 (1), divides into the first number of 38 (3) three times. Guess 3.

Step 2—Multiply: $3 \times 12 = 36$.

Step 3—Subtract: $38 - 36 = 2$.

Step 4—There are no more numbers to bring down. The remainder is 2. Write r 2 beside the 3.

Check:

	Step 1	Step 2
$\begin{array}{r}3\ r\ 2\\12\overline{)38}\end{array}$	$\begin{array}{r}12\\\times\ \ 3\\\hline36\end{array}$	$\begin{array}{r}36\\+\ \ 2\\\hline38\end{array}$

Step 1—Multiply your answer by the number you divided by: $12 \times 3 = 36$.

Step 2—Add the remainder: $36 + 2 = 38$.

Remember that the remainder must be smaller than the number you divide by. If it is not, you must start over and make a larger guess.

Look at the example below:

	Step 1	Step 2
14)76̄	$\begin{array}{r} 4\ r\ 20 \\ 14)\overline{76} \\ 56 \\ \hline 20 \end{array}$ (remainder too large)	$\begin{array}{r} 5\ r\ 6 \\ 14)\overline{76} \\ 70 \\ \hline 6 \end{array}$ (remainder OK)

Step 1—The remainder 20 is larger than 14. Try a larger guess.

Step 2—Guess the next larger number, 5. The remainder 6 is smaller than 14. The answer is correct.

Try It (Answers on page 90.)

Divide. Check your answers.

1. 22)47̄ 2. 25)59̄ 3. 16)38̄ 4. 14)43̄ 5. 12)39̄

6. 11)37̄ 7. 24)50̄ 8. 13)45̄ 9. 15)68̄ 10. 31)97̄

11. 41)89̄ 12. 38)78̄ 13. 27)86̄ 14. 14)59̄ 15. 23)72̄

GUESSING—TENS INTO HUNDREDS NUMBERS

Guessing can help you to divide larger numbers too. Look at the example below:

	Step 1	Step 2
43)129̄	$\begin{array}{r} 3 \\ 43)\overline{129} \end{array}$	$\begin{array}{r} 3 \\ 43)\overline{129} \\ -\ 129 \\ \hline 000 \end{array}$

Step 1—The first number of 43 (4) will divide into the first two numbers of 129 (12) three times. $12 \div 4 = 3$. Guess that $129 \div 43 = 3$.

Step 2—Check your guess by multiplying: $3 \times 43 = 129$. There is no remainder: $129 - 129 = 0$.

Try It (Answers on page 90.)

1. $36\overline{)144}$ 2. $54\overline{)270}$ 3. $40\overline{)160}$ 4. $43\overline{)344}$ 5. $24\overline{)120}$

6. $63\overline{)189}$ 7. $57\overline{)350}$ 8. $34\overline{)272}$ 9. $74\overline{)299}$ 10. $64\overline{)325}$

11. $72\overline{)441}$ 12. $81\overline{)733}$ 13. $61\overline{)250}$ 14. $44\overline{)267}$ 15. $35\overline{)245}$

DIVIDING TWICE BY TENS NUMBERS

Sometimes you must divide twice. Look at this example:

	Step 1	Step 2	Step 3	Step 4
$32\overline{)992}$	$32\overline{)\overset{3}{992}}$	$32\overline{)\overset{3}{992}}$ $\underline{-96}$ 3	$32\overline{)\overset{3}{992}}$ $\underline{-96}$ 32	$32\overline{)\overset{31}{992}}$ $\underline{-96}$ 32 $\underline{-32}$ 00

Step 1—Ask yourself: How many times does 32 divide into 99? $99 \div 32 = 3$.

Step 2—Multiply: $3 \times 32 = 96$. Write 96 under 99. Subtract: $99 - 96 = 3$. The remainder is 3.

Step 3—Bring down the 2. Write it next to the remainder 3, making 32.

Step 4—Divide: $32 \div 32 = 1$. Multiply: $1 \times 32 = 32$. Subtract: $32 - 32 = 0$. There is no remainder.

Try It (Answers on page 91.)

Divide. Check your answers.

1. $35\overline{)770}$ 2. $68\overline{)884}$ 3. $27\overline{)918}$ 4. $18\overline{)756}$

5. $45\overline{)945}$ 6. $26\overline{)809}$ 7. $37\overline{)759}$ 8. $15\overline{)653}$

9. $23\overline{)487}$ 10. $52\overline{)841}$ 11. $17\overline{)731}$ 12. $18\overline{)504}$

DIVIDING SEVERAL TIMES BY TENS NUMBERS

You can use the same steps you just used for any problem. It doesn't matter how many times you must divide. Look at the example below:

	Step 1	Step 2	Step 3
$23\overline{)4,899}$	$23\overline{)4899}$ 2 $\underline{-46}$ 2	$23\overline{)4899}$ 21 $\underline{-46}$ 29 $\underline{-23}$ 6	$23\overline{)4899}$ 213 $\underline{-46}$ 29 $\underline{-23}$ 69 $\underline{-69}$ 00

Step 1—Divide: $48 \div 23 = 2$. Check by multiplying: $2 \times 23 = 46$.
Subtract: $48 - 46 = 2$. There is a remainder of 2.

Step 2—Bring down the 9, making 29. Divide: $29 \div 23 = 1$. Multiply: $23 \times 1 = 23$. Subtract: $29 - 23 = 6$. There is a remainder of 6.

Step 3—Bring down the 9, making 69. Divide: $69 \div 23 = 3$. Multiply: $23 \times 3 = 69$. Subtract: $69 - 69 = 0$. There is no remainder.

Try It (Answers on page 91.)

Divide. Check your answers.

1. $34\overline{)3{,}808}$ 2. $25\overline{)5{,}776}$ 3. $42\overline{)8{,}867}$ 4. $19\overline{)6{,}099}$

5. $54\overline{)71{,}982}$ 6. $13\overline{)5{,}395}$ 7. $28\overline{)14{,}056}$ 8. $33\overline{)10{,}197}$

DIVIDING BY HUNDREDS NUMBERS

You must make a guess to divide the problem below.

	Step 1	Step 2
$234\overline{)478}$	$234\overline{)478}^{2}$	$\begin{array}{r} 2\,r\,10 \\ 234\overline{)478} \\ -\,468 \\ \hline 10 \end{array}$

Step 1—The first number of 234 (2) divides into the first number of 478 (4) two times. Guess that $478 \div 234 = 2$.

Step 2—Check by multiplying: $2 \times 234 = 468$. Subtract: $478 - 468 = 10$. There is a remainder of 10.

Try It (Answers on page 91.)

Divide. Check your answers.

1. $225\overline{)698}$ 2. $323\overline{)974}$ 3. $140\overline{)564}$ 4. $365\overline{)735}$

5. $467\overline{)998}$ 6. $502\overline{)1{,}526}$ 7. $289\overline{)1{,}256}$ 8. $176\overline{)900}$

DIVIDING TWO OR MORE TIMES BY HUNDREDS NUMBERS

Sometimes you must divide two or more times. Look at the example on the next page:

	Step 1	Step 2
325)6,847	2 325)6847 − 650 34	21 r 22 325)6847 − 650 347 − 325 22

Step 1—Divide: 684 ÷ 325 = 2. Check by multiplying 2 × 325 = 650. Write 650 under 684 and subtract. There is a remainder of 34.

Step 2—Bring down the 7. Write it next to the remainder of 34, making 347. Divide: 347 ÷ 325 = 1. Check by multiplying 1 × 325 = 325. Write 325 under 347 and subtract. There is a remainder of 22.

Try It (Answers on page 91.)

Divide. Check your answers.

1.   415)8,746   2.   232)7,428   3.   145)3,047

4.   211)88,892   5.   546)44,772   6.   312)13,420

7.   203)8,200   8.   455)10,040   9.   531)50,500

WORKING WITH WORD PROBLEMS

Read the Steps for Solving Word Problems on page xii again. Then work these problems. Find the correct answer.

1. Renee's salary is $8,448 a year. How much does she earn each month? (1 year = 12 months)
   (1) $754   (2) $704   (3) $703   (4) $600

2. George's income after taxes is $10,504 a year. How much does he earn each week? (1 year = 52 weeks)
   (1) $199   (2) $201   (3) $200   (4) $202

3. It is about 3,280 miles from New York City to Los Angeles. Jack's car gets 20 miles per gallon. How many gallons will he need for the trip?
   (1) 112   (2) 111   (3) 152   (4) 164

4. Eric's labor union collected $109,560 in dues last year. There were 415 people in the union. How much were Eric's union dues?
   (1) $415   (2) $364   (3) $264   (4) $26

5. Jack Gibson had a season record of 1,344 yards gained on 56 passes. How many yards per pass did he average?
   (1) 20   (2) 23   (3) 24   (4) 25

6. Capitol City spent $772,968 on education last year. There were 856 students enrolled in all the city schools. How much was spent on each student?
   (1) $1,003   (2) $903   (3) $803   (4) $9,033

7. Delia and Fred want to buy a used car. It is $1,332, including all credit charges. They agree to make 36 monthly payments. How much will each monthly payment be?
   (1) $37   (2) $38   (3) $39   (4) $40

8. Anne borrowed $960. She agrees to repay the loan in 24 months. How much will she pay each month?
   (1) $4   (2) $40   (3) $37   (4) $24

9. $5,544 in entry fees was collected at a stock car race. There were 132 cars competing in the race. How much was the entry fee?
   (1) $40   (2) $42   (3) $45   (4) $52

10. A subway train broke down last week. There were 375 persons on the train. The train had 15 cars. What was the average number in each car?
    (1) 20   (2) 23   (3) 25   (4) 26

Check your answers on page 91.

11. MEASURING YOUR PROGRESS

TEST A

Add

1. $\begin{array}{r} 6 \\ + 9 \\ \hline \end{array}$

2. $\begin{array}{r} 8 \\ + 6 \\ \hline \end{array}$

3. $\begin{array}{r} 5 \\ + 8 \\ \hline \end{array}$

4. $\begin{array}{r} 46 \\ + 55 \\ \hline \end{array}$

5. $\begin{array}{r} 283 \\ + 77 \\ \hline \end{array}$

6. $\begin{array}{r} 183 \\ + 379 \\ \hline \end{array}$

7. $\begin{array}{r} 6,434 \\ 2,863 \\ + 3,467 \\ \hline \end{array}$

8. $\begin{array}{r} 58,600 \\ + 19,850 \\ \hline \end{array}$

9. $\begin{array}{r} 4,067 \\ 339 \\ 842 \\ + 369 \\ \hline \end{array}$

10. $\begin{array}{r} 4,367 \\ 963 \\ 8,469 \\ + 95,701 \\ \hline \end{array}$

Subtract

11. $\begin{array}{r} 5 \\ - 2 \\ \hline \end{array}$

12. $\begin{array}{r} 16 \\ - 9 \\ \hline \end{array}$

13. $\begin{array}{r} 9 \\ - 9 \\ \hline \end{array}$

14. $\begin{array}{r} 28 \\ - 5 \\ \hline \end{array}$

15. $\begin{array}{r} 87 \\ - 68 \\ \hline \end{array}$

16. $\begin{array}{r} 465 \\ - 359 \\ \hline \end{array}$

17. $\begin{array}{r} 1,001 \\ - 99 \\ \hline \end{array}$

18. $\begin{array}{r} 3,962 \\ - 1,487 \\ \hline \end{array}$

19. $\begin{array}{r} 5,000 \\ - 3,792 \\ \hline \end{array}$

20. $\begin{array}{r} 89,463 \\ - 15,684 \\ \hline \end{array}$

Multiply

21. $\begin{array}{r} 6 \\ \times 9 \\ \hline \end{array}$

22. $\begin{array}{r} 9 \\ \times 9 \\ \hline \end{array}$

23. $\begin{array}{r} 24 \\ \times 3 \\ \hline \end{array}$

24. $\begin{array}{r} 7 \\ \times 6 \\ \hline \end{array}$

25. $\begin{array}{r} 49 \\ \times 4 \\ \hline \end{array}$

26. $\begin{array}{r} 3,694 \\ \times 2 \\ \hline \end{array}$

27. $\begin{array}{r} 37 \\ \times 45 \\ \hline \end{array}$

28. $\begin{array}{r} 337 \\ \times 17 \\ \hline \end{array}$

29. $\begin{array}{r} 467 \\ \times 40 \\ \hline \end{array}$

30. $\begin{array}{r} 8,763 \\ \times 496 \\ \hline \end{array}$

Divide

31. $6\overline{)18}$ 32. $3\overline{)159}$ 33. $9\overline{)639}$ 34. $4\overline{)928}$

35. $8\overline{)78,496}$ 36. $69\overline{)276,207}$ 37. $89\overline{)45,039}$ 38. $542\overline{)350,132}$

39. $83\overline{)25,232}$ 40. $42\overline{)126,252}$

TEST B

Read the Steps for Solving Word Problems on page xii again. Look for clue words in these problems. The clue words will help you decide whether to add, substract, multiply or divide.

1. Betty has walked three blocks. She has to walk four more blocks to the drugstore. How many blocks must she walk altogether?
   (1) 8   (2) 6   (3) 7   (4) 9

2. Laura is selling tickets for a church supper. She had 25. She has already sold 15. How many more does she have to sell?
   (1) 10   (2) 11   (3) 13   (4) 12

3. There are five people in the Thornhill family. Everyone eats two eggs for breakfast. How many eggs does the family eat altogether for breakfast?
   (1) 5   (2) 15   (3) 20   (4) 10

4. Mrs. Mills offered a $1.99 breakfast special. She cooked 27 eggs for nine customers. How many eggs did she cook for each customer?
   (1) 4   (2) 3   (3) 2   (4) 5

5. Peggy can type 35 words per minute. How many words can she type in three minutes?
   (1) 105   (2) 95   (3) 115   (4) 100

6. In last night's basketball game John made eight baskets and Warren made seven. How many baskets did the two make?
   (1) 13   (2) 15   (3) 16   (4) 14

7. Karla is making a dress. It needs 12 buttons. She ran out of buttons after she sewed on nine. How many more does she need?
   (1) 2   (2) 3   (3) 4   (4) 5

8. Mr. Kimble packed two pieces of luggage for a plane trip. One piece weighed 19 pounds. The other, 28 pounds. He is only allowed 40 pounds of luggage. How many pounds overweight was his luggage?
   (1) 8   (2) 10   (3) 7   (4) 11

9. A bushel of potatoes weighs 40 pounds. How many pounds of potatoes are there in nine bushels?
   (1) 300   (2) 460   (3) 360   (4) 380

10. Sunnifarms sells eggs in cartons of 12. How many cartons will they need for 444 eggs?
    (1) 37   (2) 47   (3) 39   (4) 57

11. Rose Ann is on a diet. How many calories are in her dinner: one slice of roast beef, 175; baked potato, 125; peas, 70; and iced tea with sugar, 150?
    (1) 520   (2) 540   (3) 525   (4) 630

12. Julio wants to pay cash for a $429 video cassette recorder. He has saved $289. How much more must he save?
    (1) $140   (2) $240   (3) $131   (4) $203

13. Mrs. Minefee went shopping at a thrift store. She bought a coat for $14, a dress for $7, and a skirt for $6. How much did she spend in all?
    (1) 27   (2) 29   (3) 28   (4) 26

14. Margie uses 27 inches of terry material to make a towel. She has 486 inches of material. How many towels can she make?
    (1) 22   (2) 16   (3) 18   (4) 15

15. During March, Mr. Booth worked 40 hours the first week; 37 hours the second; 46 hours the third, and 44 hours the fourth. How many hours did he work in March?
    (1) 157   (2) 197   (3) 167   (4) 187

16. Mr. Reed earned $7,800 last year. How much did he earn each week? (52 weeks = 1 year).
    (1) $100   (2) $200   (3) $150   (4) $250

17. A poultry farmer raised 28,763 chickens this year. He sold 27,963. Find the number of chickens that remained unsold.
    (1) 700   (2) 800   (3) 749   (4) 639

18. Harper Insurance Company employed 25 file clerks. The company paid them a total of $4,375 a week. Each clerk received the same salary. How much was each paid?
    (1) $95   (2) $115   (3) $125   (4) $175

19. Greg's rate for lawn mowing is $5 per hour. He worked six hours at Mr. Carlevatti's house. How much does Mr. Carlevatti owe Greg?
    (1) $30   (2) $25   (3) $20   (4) $15

20. Futura Furniture produced 645 pieces of furniture a week for 18 weeks. How many pieces were produced?
    (1) 10,510   (2) 11,610   (3) 13,810   (4) 11,710

21. A cup of strawberry yogurt has 260 calories. How many calories are there in three cups?
    (1) 260   (2) 780   (3) 263   (4) 257

22. Raquel and two friends drove 216 miles. They shared the driving equally. How many miles did each drive?
    (1) 72   (2) 648   (3) 213   (4) 219

23. Mr. Thompson had $178 at the beginning of the week. He has spent $89. How much does he have left?
    (1) $98   (2) $267   (3) $89   (4) $97

24. Jose weighs 184 pounds. Last year, he weighed 158 pounds. How many pounds has he gained?
    (1) 26   (2) 34   (3) 24   (4) 45

25. Cinema Mall Theater sold 154 children's tickets and 238 adult tickets. How many tickets did they sell altogether?
    (1) 124   (2) 384   (3) 182   (4) 392

Check your answers on page 92.

ANSWERS

HOW MUCH DO YOU KNOW ALREADY?, PAGE 1

Put an X by each problem you missed. Put an X by each problem you did not do. The problems are listed under the titles of the chapters in this book. If you have <u>one or more</u> Xs in a section, study that chapter to improve your skills.

ADDITION (pages 11-18)

1. **9** 2. **9** 3. **89** 4. **357** 5. **577** 6. **6** 7. **9** 8. **87** 9. **969**

MORE ON ADDITION (pages 19-24)

10. **15** 11. **135** 12. **85** 13. **70** 14. **844** 15. **8,356** 16. **9,180**
17. **4,000** 18. **25,125**

SUBTRACTION (pages 25-31)

19. **3** 20. **2** 21. **22** 22. **121** 23. **4,001** 24. **8,621**

MORE ON SUBTRACTION (pages 32-38)

25. **7,844** 26. **182** 27. **2,866** 28. **169** 29. **5,377** 30. **9,708**

MULTIPLICATION (pages 39-47)

31. **72** 32. **72** 33. **0** 34. **80** 35. **292** 36. **8,025**

MORE ON MULTIPLICATION (pages 48-52)

37. **66,504** 38. **8,903** 39. **261,120** 40. **4,265,011** 41. **24,888**
42. **7,631,334**

DIVISION (pages 53-64)

43. **9** 44. **13** 45. **30** 46. **102** 47. **41** 48. **8 r 3**

MORE ON DIVISION (pages 65-72)

49. **3 r 2** 50. **3** 51. **31** 52. **112** 53. **21 r 22** 54. **213**

TRY IT, PAGE 5

1. 42 has **4** tens and **2** ones.
2. 89 has **8** tens and **9** ones.
3. 31 has **3** tens and **1** one.
4. 55 has **5** tens and **5** ones.
5. 93 has **9** tens and **3** ones.
6. 67 has **6** tens and **7** ones.
7. 28 has **2** tens and **8** ones.
8. 16 has **1** ten and **6** ones.
9. 74 has **7** tens and **4** ones.
10. 92 has **9** tens and **2** ones.

TRY IT, PAGE 6, TOP

1. 689 has **6** hundreds, **8** tens, and **9** ones.
2. 471 has **4** hundreds, **7** tens, and **1** one.
3. 513 has **5** hundreds, **1** ten, and **3** ones.
4. 837 has **8** hundreds, **3** tens, and **7** ones.
5. 125 has **1** hundred, **2** tens, and **5** ones.
6. 952 has **9** hundreds, **5** tens, and **2** ones.
7. 264 has **2** hundreds, **6** tens, and **4** ones.
8. 398 has **3** hundreds, **9** tens, and **8** ones.
9. 612 has **6** hundreds, **1** ten, and **2** ones.
10. 746 has **7** hundreds, **4** tens, and **6** ones.

TRY IT, PAGE 6, BOTTOM

1. 1,535 has **1** thousand, **5** hundreds, **3** tens, and **5** ones.
2. 6,421 has **6** thousands, **4** hundreds, **2** tens, and **1** one.
3. 3,789 has **3** thousands, **7** hundreds, **8** tens, and **9** ones.
4. 4,618 has **4** thousands, **6** hundreds, **1** ten, and **8** ones.
5. 7,296 has **7** thousands, **2** hundreds, **9** tens, and **6** ones.
6. 5,812 has **5** thousands, **8** hundreds, **1** ten, and **2** ones.
7. 9,374 has **9** thousands, **3** hundreds, **7** tens, and **4** ones.
8. 4,965 has **4** thousands, **9** hundreds, **6** tens, and **5** ones.
9. 8,147 has **8** thousands, **1** hundred, **4** tens, and **7** ones.
10. 2,953 has **2** thousands, **9** hundreds, **5** tens, and **3** ones.

TRY IT, PAGE 7

1. **33,492** 2. **16,856** 3. **41,937** 4. **22,221**
5. **94,512** 6. **57,368** 7. **86,573**

TRY IT, PAGE 8, MIDDLE

1. **six hundred forty-eight thousand, nine hundred twenty-seven**
2. **two hundred ninety-five thousand, six hundred thirty-one**
3. **three hundred twenty-four thousand, seven hundred sixty-eight**
4. **four hundred nineteen thousand, two hundred fifty-three**
5. **one hundred sixty-three thousand, six hundred forty-seven**

TRY IT, PAGE 8, BOTTOM

1. **3,418,623** 2. **5,245,651** 3. **6,661,277**
4. **4,124,338** 5. **2,221,812** 6. **9,311,543**

TRY IT, PAGE 9

1. **5,016** 2. **11,001** 3. **50,000**
4. **906,042** 5. **43,014** 6. **1,000,000**

TRY IT, PAGE 10

1. **53,000** 2. **500** 3. **480,000**
4. **500,000** 5. **92,000**

TRY IT, PAGE 11

1. **4**   2. **9**   3. **8**   4. **6**   5. **8**
6. **5**   7. **8**   8. **4**   9. **9**   10. **7**

TRY IT, PAGE 12, TOP

1.   6
 + 3
   9
	2. 1
	 + 5
	   6
		3. 2
		 + 3
		   5
			4. 7
			 + 2
			   9

5.   3
 + 4
   7
	6. 3
	 + 3
	   6
		7. 2
		 + 1
		   3
			8. 1
			 + 8
			   9

TRY IT, PAGE 12, BOTTOM

1. **47**   2. **68**   3. **85**   4. **99**   5. **77**
6. **83**   7. **54**   8. **29**   9. **38**   10. **89**

TRY IT, PAGE 13

1. **979**   2. **785**   3. **958**   4. **897**   5. **777**
6. **699**   7. **651**   8. **859**   9. **703**   10. **469**

TRY IT, PAGE 14, TOP

1. **48**   2. **388**   3. **686**   4. **189**   5. **279**
6. **49**   7. **495**   8. **878**   9. **299**   10. **898**

TRY IT, PAGE 14, BOTTOM

1. **9**   2. **8**   3. **8**   4. **7**   5. **8**   6. **1**
7. **5**   8. **9**   9. **8**   10. **7**   11. **6**   12. **7**

TRY IT, PAGE 15

1. **8**   2. **9**   3. **9**   4. **8**   5. **6**
6. **9**   7. **9**   8. **9**   9. **8**   10. **9**

TRY IT, PAGE 16

1. **96**   2. **85**   3. **45**   4. **87**   5. **66**
6. **68**   7. **99**   8. **37**   9. **87**   10. **79**
11. **856**   12. **916**   13. **886**   14. **269**   15. **318**
16. **7,797**   17. **9,839**   18. **5,785**   19. **7,489**

WORKING WITH WORD PROBLEMS, PAGE 17

1. **(2)**
```
  2
  3
+ 1
───
  6
```

2. **(4)**
```
  2
  2
  3
+ 1
───
  8
```

3. **(3)**
```
  2
  3
+ 3
───
  8
```

4. **(3)**
```
 37
+12
───
 49
```

5. **(1)**
```
  11
   0
  21
   5
+ 10
────
  47
```

6. **(1)**
```
  10
   4
  12
+  3
────
  29
```

7. **(3)**
```
 112
 231
+315
────
 658
```

8. **(3)**
```
 421
 355
+112
────
 888
```

9. **(4)**
```
  113
  102
  141
+ 133
─────
  489
```

10. **(3)**
```
    6,324
+   6,260
─────────
 $12,584
```

TRY IT, PAGE 19, TOP

1. **10**   2. **17**   3. **13**   4. **12**   5. **11**
6. **18**   7. **15**   8. **15**   9. **13**   10. **16**

TRY IT, PAGE 19, BOTTOM

1. **116**   2. **129**   3. **188**   4. **128**   5. **159**
6. **1,378**   7. **1,666**   8. **1,046**   9. **1,143**   10. **1,489**

TRY IT, PAGE 20

1. **62**   2. **81**   3. **90**   4. **46**   **94**
6. **95**   7. **37**   8. **50**   9. **81**   10. **64**
11. **83**   12. **52**   13. **96**   14. **83**   15. **63**
16. **138**   17. **143**   18. **161**   19. **110**   20. **100**

TRY IT, PAGE 21

1. **825**	2. **739**	3. **816**	4. **529**	5. **957**
6. **700**	7. **384**	8. **939**	9. **818**	10. **909**
11. **1,479**	12. **1,427**	13. **1,229**	14. **1,968**	15. **1,406**

TRY IT, PAGE 22, TOP

1. **6,295**	2. **8,285**	3. **7,592**	4. **4,399**	5. **8,058**
6. **9,799**	7. **7,039**	8. **6,376**	9. **9,122**	10. **7,189**

TRY IT, PAGE 22, BOTTOM

1. **9,124**	2. **6,721**	3. **8,421**	4. **4,000**
5. **12,131**	6. **13,400**	7. **10,403**	8. **12,120**
9. **10,217**	10. **16,547**	11. **13,562**	12. **15,527**
13. **8,239**	14. **10,557**	15. **9,621**	16. **17,432**
17. **21,049**	18. **11,230**	19. **10,100**	20. **3,359**

WORKING WITH WORD PROBLEMS, PAGE 23

1. **(3)**

$$
\begin{array}{r}
7 \\
+\ 4 \\
\hline
\mathbf{11}
\end{array}
$$

2. **(2)**

$$
\begin{array}{r}
7 \\
4 \\
3 \\
+\ 7 \\
\hline
\mathbf{21}
\end{array}
$$

3. **(1)**

$$
\begin{array}{r}
36 \\
31 \\
+\ 40 \\
\hline
\mathbf{107}
\end{array}
$$

4. **(2)**

$$
\begin{array}{r}
166 \\
242 \\
135 \\
+\ 154 \\
\hline
\mathbf{697}
\end{array}
$$

5. **(2)**

$$
\begin{array}{r}
92 \\
126 \\
85 \\
+\ 102 \\
\hline
\mathbf{405}
\end{array}
$$

6. **(3)**

$$
\begin{array}{r}
145 \\
273 \\
198 \\
205 \\
+\ 210 \\
\hline
\mathbf{1,031}
\end{array}
$$

7. **(1)**

$$
\begin{array}{r}
1,263 \\
1,725 \\
+\ 876 \\
\hline
\mathbf{3,864}
\end{array}
$$

8. **(2)**

$$
\begin{array}{r}
947 \\
1,093 \\
756 \\
1,100 \\
+\ 1,000 \\
\hline
\mathbf{4,896}
\end{array}
$$

9. **(4)**

$$
\begin{array}{r}
2,550 \\
1,437 \\
880 \\
+\ 643 \\
\hline
\mathbf{5,510}
\end{array}
$$

10. **(4)**

$$
\begin{array}{r}
40,963 \\
37,461 \\
40,000 \\
34,312 \\
+\ 50,023 \\
\hline
\mathbf{202,759}
\end{array}
$$

TRY IT, PAGE 25

1. **1**	2. **2**	3. **4**	4. **6**	5. **2**	6. **1**
7. **4**	8. **1**	9. **3**	10. **2**	11. **3**	12. **2**

TRY IT, PAGE 26

1.
$$\begin{array}{r} 7 \\ -3 \\ \hline 4 \end{array}$$

2.
$$\begin{array}{r} 9 \\ -4 \\ \hline 5 \end{array}$$

3.
$$\begin{array}{r} 3 \\ -1 \\ \hline 2 \end{array}$$

4.
$$\begin{array}{r} 6 \\ -5 \\ \hline 1 \end{array}$$

5.
$$\begin{array}{r} 8 \\ -3 \\ \hline 5 \end{array}$$

6.
$$\begin{array}{r} 5 \\ -2 \\ \hline 3 \end{array}$$

7.
$$\begin{array}{r} 9 \\ -7 \\ \hline 2 \end{array}$$

8.
$$\begin{array}{r} 8 \\ -2 \\ \hline 6 \end{array}$$

TRY IT, PAGE 27

1. **13**	2. **14**	3. **35**	4. **22**	5. **21**
6. **35**	7. **14**	8. **34**	9. **31**	10. **45**
11. **13**	12. **24**	13. **51**	14. **32**	15. **26**

TRY IT, PAGE 28, TOP

1. **212**	2. **435**	3. **133**	4. **343**	5. **513**
6. **1,242**	7. **2,431**	8. **2,201**	9. **2,125**	10. **1,022**
11. **35,111**	12. **72,361**	13. **52,310**	14. **10,204**	15. **23,413**

TRY IT, PAGE 28, BOTTOM

1. **10**	2. **100**	3. **107**	4. **410**	5. **402**
6. **2,024**	7. **7,002**	8. **2,000**	9. **2,041**	10. **2,010**

TRY IT, PAGE 29

1. **33**	2. **83**	3. **23**	4. **163**	5. **861**
6. **2,744**	7. **7,631**	8. **8,592**	9. **3,531**	10. **1,032**

WORKING WITH WORD PROBLEMS, PAGE 30

1. **(4)**   9
             − 5
             —
              4

2. **(2)**   50
             − 26
             —
              24

3. **(1)**   193
             − 51
             —
              142

4. **(1)**   147
             − 15
             —
              132

5. **(4)**   466
              356
             —
              110

6. **(1)**   19
             − 9
             —
              10

7. **(2)**   60
             − 30
             —
              30

8. **(4)**   97,530
             − 56,430
             —
              41,100

9. **(3)**   42
             − 12
             —
              30

10. **(4)**   1,908
              − 805
              —
               1,103

TRY IT, PAGE 32

1. **9**   2. **7**   3. **9**   4. **8**   5. **8**
6. **6**   7. **8**   8. **9**   9. **9**   10. **7**

TRY IT, PAGE 33

1. **24**   2. **48**   3. **19**   4. **27**   5. **8**
6. **38**   7. **15**   8. **29**   9. **17**   10. **19**
11. **38**   12. **9**   13. **37**   14. **29**   15. **48**
16. **17**   17. **29**   18. **16**   19. **39**   20. **56**

TRY IT, PAGE 34

1. **582**   2. **481**   3. **282**   4. **491**   5. **293**
6. **183**   7. **163**   8. **394**   9. **81**   10. **293**
11. **581**   12. **261**   13. **73**   14. **155**   15. **382**

TRY IT, PAGE 35

1. **5,921**   2. **1,832**   3. **3,732**   4. **3,714**
5. **3,473**   6. **1,721**   7. **4,821**   8. **1,916**
9. **7,933**   10. **1,733**   11. **6,415**   12. **8,956**

TRY IT, PAGE 36, TOP

1. **176**   2. **189**   3. **189**   4. **567**   5. **158**
6. **868**   7. **1,878** 8. **577**   9. **2,891** 10. **1,846**
11. **7,045** 12. **3,903** 13. **8.876** 14. **6,133** 15. **2,565**

TRY IT, PAGE 36, BOTTOM

1. **649**   2. **169**   3. **139**   4. **159**   5. **377**
6. **5,238** 7. **2,192** 8. **3,656** 9. **2,087** 10. **3,657**

TRY IT, PAGE 37

1. **2,609** 2. **32**   3. **2,369** 4. **321**   5. **1,279**
6. **8,005** 7. **9,369** 8. **44,927** 9. **77,476** 10. **37,977**

WORKING WITH WORD PROBLEMS, PAGE 38

1. **(2)**   11
     −   5
     6

2. **(4)**   180
     − 165
     15

3. **(2)**   38
     − 29
     9

4. **(3)**   18,421
     − 10,638
     7,783

5. **(2)**   540
     − 145
     395

6. **(2)**   1,000
     −   896
     104

7. **(3)**   423
     −   85
     338

8. **(4)**   31,500
     − 26,489
     5,011

9. **(1)**   99,000
     − 75,998
     23,002

10. **(2)**   51,246
     − 31,484
     19,762

TRY IT, PAGE 40

<u>A</u>	<u>B</u>
1. **24**	1. **24**
2. **72**	2. **72**
3. **35**	3. **35**
4. **18**	4. **18**
5. **14**	5. **14**

TRY IT, PAGE 41, TOP

1. **0**   2. **0**   3. **0**   4. **0**   5. **0**

TRY IT, PAGE 41, BOTTOM

1. **9**   2. **7**   3. **3**   4. **8**   5. **6**

TRY IT, PAGE 42, TOP

1. **39**	2. **42**	3. **24**	4. **60**	5. **77**
6. **69**	7. **82**	8. **88**	9. **98**	10. **28**
11. **80**	12. **93**	13. **86**	14. **36**	15. **66**

TRY IT, PAGE 42, BOTTOM

1. **72**	2. **65**	3. **74**	4. **92**	5. **84**
6. **64**	7. **84**	8. **75**	9. **78**	10. **68**
11. **54**	12. **78**	13. **74**	14. **76**	15. **72**

TRY IT, PAGE 43

1. **939**	2. **693**	3. **286**	4. **606**	5. **846**
6. **642**	7. **690**	8. **663**	9. **666**	10. **408**

TRY IT, PAGE 44

1. **668**	2. **858**	3. **990**	4. **948**	5. **925**
6. **921**	7. **968**	8. **813**	9. **278**	10. **740**
11. **1,820**	12. **5,467**	13. **5,112**	14. **5,862**	15. **3,672**

TRY IT, PAGE 45, TOP

1. **6,042**	2. **2,648**	3. **9,096**	4. **4,880**	5. **3,306**
6. **4,206**	7. **48,080**	8. **6,826**	9. **4,662**	10. **96,690**

TRY IT, PAGE 45, BOTTOM

1. **5,388**	2. **8,061**	3. **7,812**	4. **8,553**	5. **7,827**
6. **9,944**	7. **49,158**	8. **64,740**	9. **7,140**	10. **8,151**
11. **33,872**	12. **70,290**	13. **28,045**	14. **18,252**	15. **20,972**

WORKING WITH WORD PROBLEMS, PAGE 46

1. **(3)**
$$\begin{array}{r} 9 \\ \times\ 6 \\ \hline \mathbf{54} \end{array}$$

2. **(4)**
$$\begin{array}{r} 28 \\ \times\ 6 \\ \hline \mathbf{168} \end{array}$$

3. **(1)**
$$\begin{array}{r} 17 \\ \times\ 9 \\ \hline \mathbf{153} \end{array}$$

4. **(2)**
$$\begin{array}{r} 350 \\ \times\ 7 \\ \hline \mathbf{2,450} \end{array}$$

5. **(2)**
$$\begin{array}{r} 1,487 \\ \times\ 8 \\ \hline \mathbf{11,896} \end{array}$$

6. **(3)**
$$\begin{array}{r} 18 \\ \times\ 8 \\ \hline 144 \end{array} \qquad \begin{array}{r} 144 \\ \times\ 4 \\ \hline \mathbf{576} \end{array}$$

7. **(1)**
$$\begin{array}{r} 559 \\ \times\ 2 \\ \hline 1,118 \end{array} \qquad \begin{array}{r} 1,118 \\ \times\ 6 \\ \hline \mathbf{6,708} \end{array}$$

8. **(3)**
$$\begin{array}{r} 456 \\ \times\ 3 \\ \hline 1,368 \end{array} \qquad \begin{array}{r} 1,368 \\ \times\ 7 \\ \hline \mathbf{9,576} \end{array}$$

9. **(3)**
$$\begin{array}{r} 45 \\ \times\ 6 \\ \hline \mathbf{\$270} \end{array}$$

10. **(3)**
$$\begin{array}{r} 12 \\ \times\ 12 \\ \hline \mathbf{144} \end{array}$$

TRY IT, PAGE 49

1. **504**	2. **759**	3. **420**	4. **880**	5. **352**
6. **2,312**	7. **1,215**	8. **2,425**	9. **25,707**	10. **66,504**
11. **278,775**	12. **46,574**	13. **23,100**	14. **160,002**	15. **675,906**
16. **30,172**	17. **224,469**	18. **318,115**	19. **20,306**	20. **75,354**
21. **1,014,447**	22. **380,160**	23. **65,307**	24. **340,684**	25. **292,400**

TRY IT, PAGE 50

1. **66,568**	2. **69,394**	3. **17,182**	4. **103,683**	5. **79,560**
6. **303,744**	7. **277,332**	8. **265,515**	9. **535,027**	10. **96,302**
11. **2,918,952**	12. **4,448,398**	13. **5,325,738**	14. **4,404,981**	15. **4,265,011**
16. **1,927,200**	17. **1,152,606**	18. **6,333,270**	19. **2,914,850**	20. **553,128**
21. **1,274,405**	22. **962,000**	23. **1,972,863**	24. **2,842,400**	25. **2,120,946**

WORKING WITH WORD PROBLEMS, PAGE 51

1. **(4)**
```
      56
×     25
     280
+ 1 120
   1,400
```

2. **(4)**
```
      45
×     26
     270
+    900
  $1,170
```

3. **(1)**
```
     215
×     24
     860
+ 4 300
  $5,160
```

4. **(3)**
```
     290
      52
     580
+ 14 500
  $15,080
```

5. **(2)**
```
     250
×     35
   1 250
+ 7 500
   8,750
```

6. **(3)**
```
   4,987
×     48
  39 896
+ 199 480
 239,376
```

7. **(4)**
```
      60        1,440
×     24    ×    365
     240      7 200
   1 20      86 400
   1,440    432 000
            525,600
```

8. **(3)**
```
     145
×     24
     580
   2 90
   3,480
```

9. **(2)**
```
      36       468
×     13    − 395
     108      $73
+   360
     468
```

10. **(4)**
```
      25         2 394
×      5     ×     125
     125       11 970
               47 880
             + 239 400
               299,250
```

TRY IT, PAGE 53

1. **4**	2. **7**	3. **3**	4. **8**	5. **5**
6. **6**	7. **9**	8. **4**	9. **6**	10. **5**
11. **9**	12. **5**	13. **5**	14. **7**	15. **3**
16. **7**	17. **6**	18. **8**	19. **7**	20. **8**

TRY IT, PAGE 55, TOP

1. **12**	2. **42**	3. **24**	4. **21**	5. **13**
6. **21**	7. **31**	8. **32**	9. **11**	10. **11**

TRY IT, PAGE 55, BOTTOM

1. **10**	2. **20**	3. **20**	4. **10**	5. **10**
6. **30**	7. **20**	8. **10**	9. **40**	10. **30**

TRY IT, PAGE 56

1. **211**	2. **122**	3. **234**	4. **121**	5. **212**
6. **332**	7. **111**	8. **431**	9. **112**	10. **211**
11. **313**	12. **342**			

TRY IT, PAGE 57

1. **201**	2. **103**	3. **203**	4. **101**	5. **301**
6. **301**	7. **102**	8. **201**	9. **101**	10. **203**

TRY IT, PAGE 58

1. **3,123**	2. **1,202**	3. **3,214**
4. **1,101**	5. **1,331**	6. **2,121**
7. **2,213**	8. **1,111**	9. **2,112**

TRY IT, PAGE 59, TOP

1. **42**	2. **51**	3. **81**	4. **82**	5. **71**
6. **91**	7. **71**	8. **73**	9. **61**	10. **54**

TRY IT, PAGE 59, BOTTOM

1. **204**	2. **103**	3. **103**	4. **107**	5. **306**
6. **103**	7. **407**	8. **106**	9. **208**	10. **105**

TRY IT, PAGE 60

1. **2r1**	2. **2r1**	3. **3r1**	4. **1r2**	5. **2r1**
6. **2r2**	7. **3r2**	8. **5r2**	9. **7r3**	10. **6r3**
11. **7r2**	12. **9r4**	13. **8r1**	14. **8r4**	15. **9r4**
16. **7r1**	17. **8r3**	18. **4r4**	19. **6r2**	20. **6r2**

TRY IT, PAGE 61

1. **16**	2. **13**	3. **12**	4. **16**	5. **15**
6. **12**	7. **14**	8. **18r1**	9. **17**	10. **13**
11. **14r4**	12. **13**	13. **24**	14. **19r1**	15. **18**

TRY IT, PAGE 63, TOP

1. **241**	2. **277**	3. **243**	4. **124**	5. **121r2**
6. **499**	7. **114**	8. **258r1**	9. **62r1**	10. **356**
11. **974**	12. **942**	13. **433**	14. **856**	15. **3,556**
16. **853**	17. **1,276**	18. **943**	19. **636r2**	20. **1,807**

WORKING WITH WORD PROBLEMS, PAGE 64

1. **(2)** $4\overline{)8}$ with quotient **2**

2. **(4)** $3\overline{)99}$ with quotient **33**

3. **(2)** $2\overline{)202}$ with quotient **101**

4. **(2)** $2\overline{)20}$ with quotient **10**

5. **(2)** $2\overline{)30}$ with quotient **15**

6. **(4)** $8\overline{)240}$ with quotient **30**

7. **(2)** $3\overline{)3,246}$ with quotient **1,082**

8. **(1)** $\begin{array}{r} 200 \\ -\ 40 \\ \hline 160 \end{array}$ $8\overline{)160}$ with quotient **20**

9. **(3)** $7\overline{)1,561}$ with quotient **223**

10. **(2)** $9\overline{)63,027}$ with quotient **7,003**

TRY IT, PAGE 65

1. **6**	2. **2**	3. **3**	4. **3**	5. **4**
6. **3**	7. **3**	8. **3**	9. **3**	10. **4**
11. **3**	12. **5**	13. **6**	14. **2**	15. **3**

TRY IT, PAGE 67

1. **2r3**	2. **2r9**	3. **2r6**	4. **3r1**	5. **3r3**
6. **3r4**	7. **2r2**	8. **3r6**	9. **4r8**	10. **3r4**
11. **2r7**	12. **2r2**	13. **3r5**	14. **4r3**	15. **3r3**

TRY IT, PAGE 68

1. **4**	2. **5**	3. **4**	4. **8**	5. **5**
6. **3**	7. **6r8**	8. **8**	9. **4r3**	10. **5r5**
11. **6r9**	12. **9r4**	13. **4r6**	14. **6r3**	15. **7**

TRY IT, PAGE 69, TOP

1. **22**   2. **13**   3. **34**   4. **42**
5. **21**   6. **31 r 3**   7. **20 r 19** 8. **43 r 8**
9. **21 r 4** 10. **16 r 9** 11. **43**   12. **28**

TRY IT, PAGE 70, TOP

1. **112**   2. **231 r 1** 3. **211 r 5** 4. **321**
5. **1,333** 6. **415**   7. **502**   8. **309**

TRY IT, PAGE 70, BOTTOM

1. **3 r 23** 2. **3 r 5**   3. **4 r 4**   4. **2 r 5**
5. **2 r 64** 6. **3 r 20** 7. **4 r 100** 8. **5 r 20**

TRY IT, PAGE 71

1. **21 r 31**   2. **32 r 4**   3. **21 r 2**
4. **421 r 61** 5. **82**   6. **43 r 4**
7. **40 r 80**   8. **22 r 30**   9. **95 r 55**

WORKING WITH WORD PROBLEMS, PAGE 72

1. **(2)** $12\overline{)8448}$ = **$704**

2. **(4)** $52\overline{)10,504}$ = **$202**

3. **(4)** $20\overline{)3,280}$ = **164**

4. **(3)** $415\overline{)109,560}$ = **$264**

5. **(3)** $56\overline{)1,344}$ = **24**

6. **(2)** $856\overline{)772,968}$ = **903**

7. **(1)** $36\overline{)1,332}$ = **$37**

8. **(2)** $24\overline{)960}$ = **$40**

9. **(2)** $132\overline{)5,544}$ = **42**

10. **(3)** $15\overline{)375}$ = **25**

MEASURING YOUR PROGRESS, PAGE 73

TEST A

ADDITION	SUBTRACTION	MULTIPLICATION	DIVISION
1. **15**	11. **3**	21. **54**	31. **3**
2. **14**	12. **7**	22. **81**	32. **53**
3. **13**	13. **0**	23. **72**	33. **71**
4. **101**	14. **23**	24. **42**	34. **232**
5. **360**	15. **19**	25. **196**	35. **9,812**
6. **562**	16. **106**	26. **7,388**	36. **4,003**
7. **12,764**	17. **902**	27. **1,665**	37. **506 r 5**
8. **78,450**	18. **2,475**	28. **5,729**	38. **646**
9. **5,617**	19. **1,208**	29. **18,680**	39. **304**
10. **109,500**	20. **73,779**	30. **4,346,448**	40. **3,006**

TEST B

1. **(3)**
```
   3
 + 4
   7
```

2. **(1)**
```
   25
 - 15
   10
```

3. **(4)**
```
   5
 × 2
  10
```

4. **(2)**
```
    3
 9)27
   27
   00
```

5. **(1)**
```
   35
 ×  3
  105
```

6. **(2)**
```
   8
 + 7
  15
```

7. **(2)**
```
   12
 -  9
    3
```

8. **(3)**
```
   19        47
 + 28      - 40
   47         7
```

9. **(3)**
```
    40
 ×   9
   360
```

10. **(1)**
```
     37
 12)444
    36
    84
    84
    00
```

11. **(1)**
```
    175
    125
     70
 + 150
    520
```

12. **(1)**
```
    429
 -  289
  $140
```

13. **(1)**
```
   14
    7
 +  6
   27
```

14. **(3)**
```
     18
 27)486
    27
    216
    216
    00
```

15. **(3)**   40
                   37
                   46
               +   44
                  167

16. **(3)**   **150**
            52)7,800
                5 2
                2 60
                2 60
                0 000

17. **(2)**   28,763
              − 27,963
                  800

18. **(4)**   **$175**
            25)4,375
               2 5
               1 87
               1 75
                125
                125
                000

19. **(1)**   5
                 × 6
                 $30

20. **(2)**   645
              ×   18
                 5 160
                 6 450
               11,610

21. **(2)**   260
              ×   3
               780

22. **(1)**   **72**
            3)216
              21
               6
               6

23. **(3)**   178
              −   89
               $ 89

24. **(1)**   184
              − 158
                 26

25. **(4)**   154
              + 238
                392

Fractions

12. HOW MUCH DO YOU KNOW ALREADY?

Here are some problems with fractions. You will be asked to add, subtract, multiply, or divide. This will help you find out how much you know already. You will also find out what you need to study. Do as many problems as you can. Write your answers on the lines at the right. Then, turn to page 133. Check your answers. Page 133 will tell you what pages in the book to study.

1. $\frac{2}{8} + \frac{1}{8} =$ 2. $\frac{2}{11} + \frac{4}{11} + \frac{3}{11} =$ 3. $\frac{1}{2} + \frac{1}{2} =$

4. $\begin{array}{r} 1\frac{3}{6} \\ + \ 4\frac{2}{6} \\ \hline \end{array}$ 5. $\begin{array}{r} 3\frac{3}{8} \\ + \ 4\frac{5}{8} \\ \hline \end{array}$ 6. $\begin{array}{r} 3\frac{4}{5} \\ + \ 2\frac{3}{5} \\ \hline \end{array}$

7. $\frac{5}{8} - \frac{2}{8} =$ 8. $\frac{7}{12} - \frac{3}{12} =$ 9. $\begin{array}{r} 3 \\ - \ \frac{3}{4} \\ \hline \end{array}$

10. $\begin{array}{r} 4\frac{7}{9} \\ - \ \frac{2}{9} \\ \hline \end{array}$ 11. $\begin{array}{r} 7\frac{1}{3} \\ - \ 2\frac{2}{3} \\ \hline \end{array}$ 12. $\begin{array}{r} 9\frac{2}{9} \\ - \ 3\frac{5}{9} \\ \hline \end{array}$

13. $\frac{1}{2} + \frac{1}{4} + \frac{1}{8} =$ 14. $2\frac{3}{4} + 4\frac{1}{2} =$ 15. $\frac{2}{3} - \frac{1}{2} =$

16. $\frac{5}{12} - \frac{1}{6} =$ 17. $\frac{2}{3} + \frac{3}{5} =$ 18. $3\frac{1}{3} - 2\frac{3}{4} =$

19. $\frac{1}{3} \times \frac{1}{3} =$ 20. $\frac{3}{4} \times \frac{2}{3} =$ 21. $2\frac{1}{2} \times \frac{3}{4} =$

22. $\frac{2}{3} \times 3\frac{1}{4} =$ 23. $\frac{2}{3} \times \frac{6}{7} =$ 24. $2\frac{2}{3} \times 3\frac{1}{2} =$

25. $\frac{1}{3} \div \frac{1}{3} =$ 26. $2 \div \frac{1}{2} =$ 27. $\frac{5}{6} \div 3\frac{1}{8} =$

28. $3\frac{1}{3} \div \frac{2}{3} =$ 29. $\frac{1}{5} \div \frac{1}{10} =$ 30. $2\frac{2}{3} \div 3\frac{1}{2} =$

1. _____
2. _____
3. _____
4. _____
5. _____
6. _____
7. _____
8. _____
9. _____
10. _____
11. _____
12. _____
13. _____
14. _____
15. _____
16. _____
17. _____
18. _____
19. _____
20. _____
21. _____
22. _____
23. _____
24. _____
25. _____
26. _____
27. _____
28. _____
29. _____
30. _____

13. WHAT IS A FRACTION?

Is this glass $\frac{1}{2}$ empty or $\frac{1}{2}$ full?

It does not matter. You may say the glass is $\frac{1}{2}$ empty or $\frac{1}{2}$ full. You know there is not a <u>whole</u> glass of water. There is only a <u>fraction</u> of a glass. A fraction is a part. 1, 2, 3, 4, 5, 6, 7, 8, and 9 are whole numbers. Fractions are not whole numbers. They are <u>parts</u> of a whole. $\frac{1}{2}$, $\frac{3}{8}$, $\frac{5}{16}$, and $\frac{3}{144}$ are fractions.

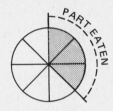

Chris and 3 friends bought this pizza. They cut it into 8 equal slices. Chris ate 3 slices. He ate 3 out of 8, or $\frac{3}{8}$, of the pizza.

$$\frac{3 = \text{the number of slices Chris ate}}{8 = \text{the total number of slices}}$$

The top number of a fraction is called the <u>numerator</u>. The numerator is the number of parts you are talking about.

The bottom number of a fraction is called the <u>denominator</u>. The denominator is the number of parts the whole is divided into.

Look at the pizza again. If $\frac{3}{8}$ of the pie was eaten:

$$\frac{3 = \text{numerator} = \text{number of slices Chris ate}}{8 = \text{denominator} = \text{number of slices the pie was divided into}}$$

Try It (Answers on page 133.)

Each pizza on the next page is cut into a different number of slices. In the first blank, write the fraction for the shaded part of each pizza. In the second blank, write the fraction for the unshaded part. The first one is done for you.

1.

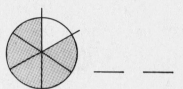

2.

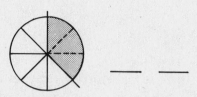

3.

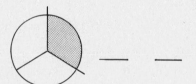

4.

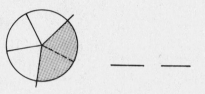

5.

6.

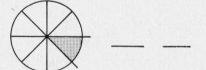

Try It Again (Answers on page 133.)

Mrs. Swados has made six boxes of fudge. They are divided differently. The dark areas have been eaten. In the first blank write the fraction for the part eaten. In the second blank write the fraction for the part left.

1.

2.

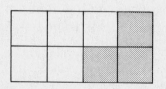

3.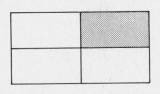

    _____

4.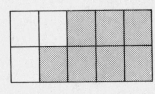

    _____   _____

5.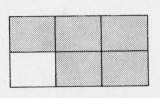

    _____   _____

6.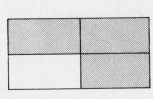

    _____   _____

PROPER AND IMPROPER FRACTIONS

The fractions you have been working with are proper fractions. Proper fractions have a top number that is smaller than the bottom number. $\frac{1}{2}$, $\frac{3}{4}$, and $\frac{2}{3}$ are proper fractions. Their value is less than 1 whole.

    Some fractions are not proper. How many $\frac{1}{2}$ circles are pictured here?

There are three $\frac{1}{2}$ circles pictured. This can also be written $\frac{3}{2}$. $\frac{3}{2}$ is an improper fraction. Improper fractions have a top number that is larger than the bottom number. Improper fractions are greater than 1 whole.

Try It (Answers on page 133.)

Circle the improper fraction in each pair.

1. $\frac{3}{4}$ or $\frac{4}{3}$

2. $\frac{23}{17}$ or $\frac{36}{40}$

3. $\frac{9}{5}$ or $\frac{4}{5}$

4. $\frac{18}{15}$ or $\frac{8}{9}$

5. $\frac{6}{3}$ or $\frac{6}{7}$

6. $\frac{6}{10}$ or $\frac{7}{3}$

7. $\frac{7}{15}$ or $\frac{6}{2}$ 8. $\frac{20}{13}$ or $\frac{16}{19}$

9. $\frac{10}{12}$ or $\frac{13}{8}$ 10. $\frac{3}{8}$ or $\frac{3}{2}$

CHANGING IMPROPER FRACTIONS TO MIXED NUMBERS

The number $1\frac{1}{2}$ is made up of two parts:

$$1\text{—a whole number}$$
$$\tfrac{1}{2}\text{—a fraction}$$

A whole number and a fraction together are called a <u>mixed number</u>. $1\frac{1}{2}$ is a mixed number. It mixes a whole number and a fraction.

The line between the top and bottom of a fraction means <u>divide</u>. The fraction $\frac{3}{2}$ means $3 \div 2$. You can change an improper fraction to a mixed number by dividing. Look at the example below:

Step 1	Step 2
$\frac{3}{2} = 2\overline{)3}^{\,1\,r\,1}$	$2\overline{)3}^{\,1\,r\,1} = 1\frac{1}{2}$

Step 1—Divide the top number by the bottom number: $\frac{3}{2} = 3 \div 2 = 1\,r\,1$.

Step 2—Place your remainder over the number you divided by (the bottom number) since the remainder is still to be divided: $\frac{3}{2} = 3 \div 2 = 1\frac{1}{2}$.

Try It (Answers on page 134.)

Change these improper fractions to mixed numbers. The first one is done for you.

1. $\frac{5}{2} = 5 \div 2 = 2\frac{1}{2}$ 2. $\frac{19}{9}$ 3. $\frac{9}{7}$ 4. $\frac{4}{3}$

5. $\frac{7}{4}$ 6. $\frac{25}{6}$ 7. $\frac{12}{5}$ 8. $\frac{7}{2}$

9. $\frac{15}{7}$ 10. $\frac{33}{10}$ 11. $\frac{14}{3}$ 12. $\frac{10}{7}$

13. $\frac{9}{2}$ 14. $\frac{13}{8}$ 15. $\frac{15}{8}$ 16. $\frac{24}{5}$

17. $\frac{11}{5}$ 18. $\frac{20}{11}$ 19. $\frac{21}{10}$ 20. $\frac{19}{8}$

CHANGING MIXED NUMBERS TO IMPROPER FRACTIONS

You can change a mixed number back to an improper fraction. You must multiply. Look at this example:

$$3\tfrac{1}{4} = \;?$$

Step 1—Multiply the bottom number times the whole number: $4 \times 3 = 12$.

Step 2—Add the top number to your answer: $12 + 1 = 13$.

Step 3—Place your answer over the bottom number: $\frac{13}{4}$.

$$3\tfrac{1}{4} = \tfrac{13}{4}$$

Try It (Answers on page 134.)

Change these mixed numbers to improper fractions. The first one is done for you.

1. $2\tfrac{1}{2} = \dfrac{(2 \times 2) + 1}{2} = \tfrac{5}{2}$ 2. $6\tfrac{4}{5}$ 3. $3\tfrac{1}{3}$

4. $4\tfrac{1}{4}$ 5. $12\tfrac{1}{2}$ 6. $1\tfrac{1}{2}$

7. $5\tfrac{2}{4}$ 8. $15\tfrac{2}{3}$ 9. $3\tfrac{3}{4}$

10. $3\tfrac{1}{8}$ 11. $4\tfrac{1}{9}$ 12. $5\tfrac{4}{5}$

13. $10\tfrac{2}{3}$ 14. $8\tfrac{3}{5}$ 15. $10\tfrac{3}{5}$

REDUCING FRACTIONS

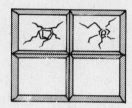

Is $\frac{2}{4}$ or $\frac{1}{2}$ of the window broken?

Both answers are correct. $\frac{2}{4}$ of the window is the same as $\frac{1}{2}$. $\frac{2}{4}$ is equal to $\frac{1}{2}$. $\frac{2}{4}$ can be renamed to its simplest form— $\frac{1}{2}$. Another way to say this is: $\frac{2}{4}$ <u>reduced to its lowest terms</u> is $\frac{1}{2}$.

To reduce fractions, divide both the top number and the bottom number by the same number. To reduce $\frac{2}{4}$ to its lowest terms, divide both the top number and the bottom number by 2:

$$\frac{2}{4} \; = \; \frac{(2 \div 2)}{(4 \div 2)} \; = \; \frac{1}{2}$$

Sometimes you can reduce a fraction by dividing both the top number and the bottom number by the top number. Look at the example below:

$$\frac{5}{15} \; = \; \frac{(5 \div 5)}{(15 \div 5)} \; = \; \frac{1}{3}$$

Sometimes the top number and the bottom number cannot be divided evenly by the top number. Try to find the largest number that will divide evenly into both the top number and the bottom number. Look at this example:

$$\frac{9}{12} \; = \; \frac{(9 \div 3)}{(12 \div 3)} \; = \; \frac{3}{4}$$

In this case, 3 is the largest number that will divide evenly into both the top and bottom numbers.

Try It (Answers on page 134.)

Reduce these fractions to lowest terms.

1. $\frac{2}{6}$

2. $\frac{12}{14}$

3. $\frac{4}{8}$

4. $\frac{6}{14}$

5. $\frac{5}{10}$

6. $\frac{4}{16}$

7. $\frac{3}{6}$

8. $\frac{4}{6}$

9. $\frac{3}{9}$

10. $\frac{6}{15}$

11. $\frac{2}{4}$

12. $\frac{10}{20}$

13. $\frac{4}{12}$

14. $\frac{12}{18}$

15. $\frac{9}{12}$

16. $\frac{10}{30}$

17. $\frac{6}{10}$

18. $\frac{7}{21}$

18. $\frac{4}{10}$

20. $\frac{20}{30}$

14. ADDING FRACTIONS

To add fractions, add only the top numbers (numerators). The bottom numbers (denominators) stay the same. Look at the example below:

$$\frac{2}{8} + \frac{1}{8} = \frac{3}{8}$$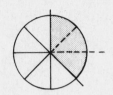

Try It (Answers on page 135.)

1. $\frac{1}{4} + \frac{2}{4} = \frac{3}{4}$

2. $\frac{7}{15} + \frac{6}{15} = \frac{}{15}$

3. $\frac{1}{20} + \frac{3}{20} + \frac{5}{20} = \frac{}{20}$

4. $\frac{6}{10} + \frac{3}{10} =$

5. $\frac{2}{6} + \frac{3}{6} =$

6. $\frac{1}{3} + \frac{1}{3} =$

7. $\frac{2}{5} + \frac{1}{5} + \frac{1}{5} =$

8. $\frac{5}{12} + \frac{4}{12} + \frac{2}{12} =$

9. $\frac{3}{7} + \frac{2}{7} =$

10. $\frac{1}{7} + \frac{2}{7} + \frac{3}{7} =$

11. $\frac{10}{19} + \frac{5}{19} + \frac{2}{19} =$

12. $\frac{1}{8} + \frac{2}{8} + \frac{2}{8} + \frac{2}{8} =$

13. $\frac{1}{8} + \frac{2}{8} + \frac{4}{8} =$

14. $\frac{3}{13} + \frac{3}{13} + \frac{3}{13} =$

15. $\frac{1}{5} + \frac{2}{5} + \frac{1}{5} =$

FRACTIONS THAT EQUAL ONE

There are four quarters in one dollar. One quarter is a fraction. It is part of a whole. It is 1 of 4, or $\frac{1}{4}$ of a dollar. Four quarters are 4 of 4 or $\frac{4}{4}$ of a dollar.

$\frac{4}{4}$ of a dollar equals 1 whole dollar.

A fraction that has the same numerator and denominator always equals 1.

Try It (Answers on page 135.)

Add. Change your answers to 1 when the top number is the same as the bottom number. Reduce to lowest terms where possible. The first one is done for you.

1. $\frac{1}{4} + \frac{3}{4} = \frac{4}{4}$, or 1

2. $\frac{2}{8} + \frac{3}{8} + \frac{3}{8} =$

3. $\frac{1}{2} + \frac{1}{2} =$

4. $\frac{4}{7} + \frac{2}{7} =$

5. $\frac{1}{5} + \frac{1}{5} + \frac{2}{5} =$

6. $\frac{13}{20} + \frac{2}{20} + \frac{3}{20} =$

7. $\frac{1}{3} + \frac{2}{3} =$

8. $\frac{1}{9} + \frac{3}{9} + \frac{3}{9} =$

9. $\frac{1}{3} + \frac{1}{3} + \frac{1}{3} =$

10. $\frac{5}{12} + \frac{5}{12} =$

11. $\frac{3}{12} + \frac{4}{12} + \frac{5}{12} =$

12. $\frac{1}{4} + \frac{1}{4} + \frac{1}{4} =$

13. $\frac{3}{8} + \frac{5}{8} =$

14. $\frac{2}{11} + \frac{4}{11} + \frac{3}{11} =$

15. $\frac{3}{8} + \frac{2}{8} + \frac{1}{8} =$

ADDING MIXED NUMBERS

You have learned to add whole numbers and fractions. Now you can add mixed numbers. Look at this example:

Step 1	Step 2	Step 3
$1\frac{1}{6}$ $+\ 4\frac{2}{6}$ $\overline{\quad \frac{3}{6}}$	$1\frac{1}{6}$ $+\ 4\frac{2}{6}$ $\overline{\quad 5\frac{3}{6}}$	$1\frac{1}{6}$ $+\ 4\frac{2}{6}$ $\overline{\quad 5\frac{3}{6} = 5\frac{1}{2}}$

Step 1—First add the fractions: $\frac{1}{6} + \frac{2}{6} = \frac{3}{6}$.

Step 2—Then add the whole numbers: $1 + 4 = 5$.

Step 3—Reduce if possible: $\frac{3}{6} = \dfrac{3 \div 3}{6 \div 3} = \frac{1}{2}$.

Try It (Answers on page 135.)

Reduce where possible.

1. $3\frac{1}{3}$
    $+\ 4\frac{1}{3}$

2. $10\frac{1}{6}$
    $+\ 1\frac{1}{6}$

3. $6\frac{2}{5}$
    $+\ 4\frac{1}{5}$

4. $7\frac{4}{9}$
    $+\ 6\frac{1}{9}$

5. $30\frac{1}{4}$
    $+\ 10\frac{1}{4}$

6. $2\frac{1}{12}$
    $+\ 3\frac{5}{12}$

7. $8\frac{4}{7}$
    $+\ \frac{2}{7}$

8. $7\frac{2}{9}$
    $+\ 4\frac{4}{9}$

9. $13\frac{2}{10}$
    $+\ 8\frac{3}{10}$

10. $19\frac{5}{7}$
    $+\ 11\frac{1}{7}$

If the numerator is the same as the denominator, add 1 to the whole number. Look at this example:

Step 1	Step 2	Step 3
$3\frac{1}{4}$ $+\ 2\frac{3}{4}$ $\overline{\quad \frac{4}{4}}$	$3\frac{1}{4}$ $+\ 2\frac{3}{4}$ $\overline{\quad 5\frac{4}{4}}$	$3\frac{1}{4}$ $+\ 2\frac{3}{4}$ $\overline{\quad 5\frac{4}{4} = 5 + 1 = 6}$

Step 1—Add the fractions: $\frac{1}{4} + \frac{3}{4} = \frac{4}{4}$.

Step 2—Add the whole numbers: $3 + 2 = 5$.

Step 3—The numerator equals the denominator. $\frac{4}{4} = 1$. Add 1 to the sum of the whole numbers: $5 + 1 = 6$.

Try It (Answers on page 135.)

1. $7\frac{2}{3}$
$+\ 8\frac{1}{3}$

2. $3\frac{3}{4}$
$+\ 4\frac{1}{4}$

3. $18\frac{2}{7}$
$+\ 9\frac{5}{7}$

4. $21\frac{4}{9}$
$+\ 4\frac{5}{9}$

5. $5\frac{3}{5}$
$+\ 5\frac{2}{5}$

6. $19\frac{4}{5}$
$+\ 7\frac{1}{5}$

7. $9\frac{3}{7}$
$+\ 3\frac{4}{7}$

8. $29\frac{1}{8}$
$+\ 10\frac{7}{8}$

9. $109\frac{8}{9}$
$+\ 90\frac{1}{9}$

10. $2\frac{1}{3}$
$3\frac{1}{3}$
$+\ 4\frac{1}{3}$

Sometimes the fraction part of the answer is greater than one. Change it to a mixed number. Add the whole part of the mixed number to the whole number in the answer. Write the remaining fraction by the answer. Look at this example:

	Step 1	Step 2	Steps 3 - 4
$3\frac{3}{5}$ $+\ 4\frac{4}{5}$	$3\frac{3}{5}$ $+\ 4\frac{4}{5}$ $\frac{7}{5}$	$3\frac{3}{5}$ $+\ 4\frac{4}{5}$ $7\frac{7}{5}$	$3\frac{3}{5}$ $+\ 4\frac{4}{5}$ $7\frac{7}{5} = 7 + 1\frac{2}{5} = 8\frac{2}{5}$

Step 1—Add the fractions: $\frac{3}{5} + \frac{4}{5} = \frac{7}{5}$.

Step 2—Add the whole numbers: $3 + 4 = 7$.

Step 3—Change the improper fraction to a mixed number: $\frac{7}{5} = 1\frac{2}{5}$.

Step 4—Add the sum of the whole numbers (7) to the mixed number: $7 + 1\frac{2}{5} = 8\frac{2}{5}$.

Try It (Answers on page 136.)

Add and reduce to lowest terms.

1. $5\frac{3}{4}$
$+\ 1\frac{2}{4}$

2. $4\frac{3}{7}$
$+\ 1\frac{6}{7}$

3. $2\frac{2}{3}$
$+\ 2\frac{2}{3}$

4. $5\frac{3}{8}$
$+\ 2\frac{7}{8}$

5. $3\frac{3}{5}$
$+\ 4\frac{4}{5}$

6. $1\frac{5}{9}$
$+\ \frac{5}{9}$

7. $6\frac{5}{7}$
$+\ 1\frac{5}{7}$

8. $6\frac{7}{10}$
$+\ 2\frac{6}{10}$

9. $9\frac{5}{7}$
$+\ 2\frac{3}{7}$

10. $\frac{3}{5}$
$+\ 2\frac{3}{5}$

11. $9\frac{3}{4}$
$+\ \frac{3}{4}$

12. $\frac{3}{4}$
$\frac{1}{4}$
$+\ \frac{3}{4}$

13. $2\frac{5}{6}$
$+\ 1\frac{3}{6}$

14. $8\frac{8}{9}$
$+\ \frac{5}{9}$

15. $\frac{2}{5}$
$3\frac{1}{5}$
$+\ 6\frac{2}{5}$

WORKING WITH WORD PROBLEMS

Read the Steps for Solving Word Problems on page xii again. Then work these problems. Find the correct answer. Reduce fractions to lowest terms where possible.

1. Rebecca is making a cake. The recipe calls for $\frac{1}{3}$ cup of flour at first and $\frac{1}{3}$ of a cup of flour later. How much flour will she use altogether?
   (1) $\frac{2}{6}$ cup   (2) $\frac{1}{6}$ cup   (3) $\frac{1}{3}$ cup   (4) $\frac{2}{3}$ cup

2. Gail's apartment has five rooms. She cleaned one room on Saturday. On Sunday she cleaned two rooms. How much of her apartment has she cleaned?
   (1) $\frac{3}{10}$ (2) $\frac{3}{5}$ (3) $\frac{1}{5}$ (4) $\frac{2}{5}$

3. Tony baked a pie and cut it into eight pieces. He ate one of the pieces right away. Yesterday he ate three pieces. Today he ate another piece. How much of the pie has he eaten?
   (1) $\frac{5}{16}$ (2) $\frac{4}{8}$ (3) $\frac{5}{8}$ (4) $\frac{5}{24}$

4. $\frac{3}{20}$ of your pay is taken for income tax. $\frac{1}{20}$ is deducted for social security. $\frac{1}{20}$ is taken for union dues. How much of your pay is taken for these expenses?
   (1) $\frac{1}{5}$ (2) $\frac{1}{4}$ (3) $\frac{5}{60}$ (4) $\frac{1}{12}$

5. Irene's motorbike takes oil and gas. She adds $\frac{2}{5}$ of a quart of oil for every gallon of gas. She put in 2 gallons of gas. How many quarts of oil does she need?
   (1) $\frac{1}{10}$ (2) $\frac{4}{10}$ (3) $\frac{4}{5}$ (4) $\frac{2}{5}$

6. Charlie bought a six-pack last weekend. He drank one can on Monday. On Tuesday he drank two cans. On Wednesday he didn't drink any. On Thursday he drank two cans. What fraction of the six-pack has Charlie finished?
   (1) $\frac{5}{18}$ (2) $\frac{3}{6}$ (3) $\frac{4}{6}$ (4) $\frac{5}{6}$

7. Mrs. Nguyen bought $2\frac{1}{3}$ yards of cloth for a skirt. She bought another $2\frac{1}{3}$ yards the next week to make a blouse. How many yards of cloth did she buy altogether?
   (1) $\frac{6}{6}$ (2) $6\frac{1}{3}$ (3) $4\frac{2}{6}$ (4) $4\frac{2}{3}$

8. Helena wants to make two ties. Each tie will take $\frac{1}{2}$ yard of material. How much material must she buy?

(1) $\frac{1}{4}$ yd   (2) $\frac{2}{4}$ yd   (3) $\frac{1}{2}$ yd   (4) 1 yd

9. At a trading stamp center, Mr. Povarnik wants a toaster for $5\frac{1}{4}$ books of stamps. He also wants a coffee pot for $2\frac{3}{4}$ books. How many books does he need?

(1) 8   (2) $7\frac{3}{4}$ (3) 7   (4) $7\frac{1}{4}$

10. To get in shape, Lester ran $\frac{3}{4}$ of a mile and rested. He then ran 1 mile and rested. Later, he ran another $\frac{3}{4}$ of a mile. How many miles did he run altogether?

(1) $1\frac{4}{8}$ (2) $1\frac{3}{4}$ (3) $2\frac{1}{2}$ (4) 1

Check your answers on page 136.

15. SUBTRACTING FRACTIONS

You have $\frac{5}{8}$ of a pie. You serve $\frac{2}{8}$ of the pie. How much of the pie is left? To find out, subtract:

$$\frac{5}{8} - \frac{2}{8} = \frac{3}{8}$$

To subtract fractions, subtract only the <u>numerators</u> or top numbers. The <u>denominators</u> or bottom numbers stay the same.

Try It (Answers on page 136.)

1. $\frac{4}{5} - \frac{1}{5} = \frac{}{5}$

2. $\frac{7}{16} - \frac{4}{16} = \frac{}{16}$

3. $\frac{20}{23} - \frac{9}{23} = \frac{}{23}$

4. $\frac{4}{5} - \frac{2}{5} =$

5. $\frac{6}{7} - \frac{4}{7} =$

6. $\frac{17}{24} - \frac{4}{24} =$

7. $\frac{9}{10} - \frac{6}{10} =$

8. $\frac{9}{13} - \frac{1}{13} =$

9. $\frac{11}{12} - \frac{6}{12} =$

10. $\frac{7}{12} - \frac{2}{12} =$

11. $\frac{7}{15} - \frac{6}{15} =$

12. $\frac{15}{16} - \frac{8}{16} =$

13. $\frac{7}{8} - \frac{4}{8} =$

14. $\frac{3}{5} - \frac{2}{5} =$

15. $\frac{19}{20} - \frac{12}{20} =$

SUBTRACTING FRACTIONS FROM THE WHOLE NUMBER 1

Charlene borrowed one cup of sugar to bake some cookies. She only uses $\frac{3}{4}$ cup. How much sugar does she have left? She has $\frac{1}{4}$ cup of sugar left. The example shows how $\frac{3}{4}$ is subtracted from the whole number 1:

$$
\begin{array}{rcr}
1 & = & \frac{4}{4} \\
- \quad \frac{3}{4} & = & - \frac{3}{4} \\
\hline
& & \frac{1}{4}
\end{array}
$$

To subtract, change the whole number 1 to a fraction. Use the same top and bottom numbers for the fraction as the bottom number of the fraction you are subtracting.

Try It (Answers on page 136.)

1.
$$1 - \frac{1}{2}$$

2.
$$1 - \frac{3}{8}$$

3.
$$1 - \frac{9}{16}$$

4.
$$1 - \frac{5}{9}$$

5.
$$1 - \frac{2}{3}$$

6.
$$1 - \frac{5}{11}$$

7.
$$1 - \frac{7}{12}$$

8.
$$1 - \frac{1}{6}$$

9.
$$1 - \frac{1}{8}$$

10.
$$1 - \frac{7}{8}$$

11.
$$1 - \frac{3}{10}$$

12.
$$1 - \frac{2}{5}$$

SUBTRACTING FRACTIONS FROM WHOLE NUMBERS

To subtract a fraction from a whole number larger than 1, you must borrow. Look at this example:

	Step 1	Step 2	Step 3
$3 - \frac{3}{8}$	$2 + 1 - \frac{3}{8}$	$2\frac{8}{8} - \frac{3}{8}$	$2\frac{8}{8} - \frac{3}{8}$ $2\frac{5}{8}$

Step 1—Borrow 1 from the 3.

Step 2—Change the 1 to eighths. $1 = \frac{8}{8}$.

Step 3—Subtract. $\frac{8}{8} - \frac{3}{8} = \frac{5}{8}$. $2 - 0 = 2$.

Try It (Answers on page 137.)

1.
$$4 - \frac{5}{6}$$

2.
$$7 - \frac{1}{3}$$

3.
$$3 - \frac{1}{2}$$

4.
$$9 - \frac{3}{4}$$

5.
$$16 - \frac{9}{10}$$

6.
$$35 - \frac{8}{13}$$

7.
$$90 - \frac{1}{15}$$

8.
$$63 - \frac{5}{8}$$

9.
$$20 - \frac{7}{8}$$

10.
$$200 - \frac{1}{5}$$

11.
$$104 - \frac{3}{10}$$

12.
$$91 - \frac{3}{8}$$

SUBTRACTING MIXED NUMBERS

To subtract mixed numbers, subtract separately. First subtract the fractions. Then subtract the whole numbers. Look at this example:

$$
\begin{array}{r}
5\frac{5}{7} \\
- 1\frac{2}{7} \\
\hline
4\frac{3}{7}
\end{array}
$$

Step 1—Subtract the fractions: $\frac{5}{7} - \frac{2}{7} = \frac{3}{7}$.

Step 2—Subtract the whole numbers: $5 - 1 = 4$.

Try It (Answers on page 137.)

Reduce answers where possible.

1. $\quad 16\frac{8}{9}$ $\quad -\ 8\frac{3}{9}$

2. $\quad 37\frac{4}{5}$ $\quad -\ 25\frac{2}{5}$

3. $\quad 21\frac{6}{7}$ $\quad -\ 18\frac{4}{7}$

4. $\quad 65\frac{10}{11}$ $\quad -\ 45\frac{7}{11}$

5. $\quad 19\frac{2}{3}$ $\quad -\ 18\frac{1}{3}$

6. $\quad 3\frac{1}{2}$ $\quad -\ 2\frac{1}{2}$

7. $\quad 28\frac{5}{8}$ $\quad -\ 19\frac{1}{8}$

8. $\quad 11\frac{4}{9}$ $\quad -\ 2\frac{1}{9}$

9. $\quad 4\frac{7}{12}$ $\quad -\ \frac{1}{12}$

10. $\quad 203\frac{3}{4}$ $\quad -\ 74\frac{3}{4}$

11. $\quad 100\frac{5}{8}$ $\quad -\ 8\frac{3}{8}$

12. $\quad 14\frac{7}{10}$ $\quad -\ 6\frac{3}{10}$

SUBTRACTING MIXED NUMBERS—BORROWING

Sometimes the fraction part of a mixed number problem cannot be subtracted. You must borrow 1 from the whole number. Change the 1 to a fraction. Add this fraction to the fraction column. Then subtract. Look at this example:

	Step 1	Step 2	Step 3	Steps 4 - 5
$7\frac{1}{4}$ $-\ 2\frac{3}{4}$	$\overset{6}{\cancel{7}}\frac{1}{4} + \frac{4}{4}$ $-\ 2\frac{3}{4}$	$6\frac{5}{4}$ $-\ 2\frac{3}{4}$	$6\frac{5}{4}$ $-\ 2\frac{3}{4}$ $\frac{2}{4}$	$6\frac{5}{4}$ $-\ 2\frac{3}{4}$ $4\frac{2}{4} = 4\frac{1}{2}$

Step 1—You cannot subtract $\frac{3}{4}$ from $\frac{1}{4}$. Borrow 1 from the 7.
Change the 1 to $\frac{4}{4}$.

Step 2—Add. $\frac{4}{4}+\frac{1}{4}=\frac{5}{4}$.

Step 3—Subtract the fractions. $\frac{5}{4}-\frac{3}{4}=\frac{2}{4}$.

Step 4—Subtract the whole numbers. $6-2=4$.

Step 5—Reduce your answer. $\frac{2}{4}=\dfrac{2\div 2}{4\div 2}=\frac{1}{2}$.

Try It (Answers on page 137.)

Reduce answers where possible.

1. $6\frac{1}{3}$
    $-\ 1\frac{2}{3}$

2. $10\frac{3}{5}$
    $-\ 6\frac{4}{5}$

3. $21\frac{4}{8}$
    $-12\frac{7}{8}$

4. $19\frac{7}{10}$
    $-\ 6\frac{9}{10}$

5. $18\frac{6}{12}$
    $-\ 6\frac{7}{12}$

6. $24\frac{5}{9}$
    $-17\frac{7}{9}$

7. $247\frac{6}{11}$
    $-163\frac{8}{11}$

8. $233\frac{1}{3}$
    $-121\frac{2}{3}$

9. $23\frac{5}{12}$
    $-13\frac{7}{12}$

10. $101\frac{3}{8}$
     $-\ 9\frac{5}{8}$

11. 130
     $-\ 5\frac{1}{4}$

12. 126
     $-\ 19\frac{5}{8}$

WORKING WITH WORD PROBLEMS

Read the Steps for Solving Word Problems on page xii again. Then work these problems. Find the correct answer. Reduce fractions to lowest terms where possible.

1. Ron is baking bread. He is low on flour. The recipe calls for $1\frac{2}{3}$ cups of flour. Ron uses $\frac{1}{3}$ cup less flour. How many cups of flour did Ron use?

   (1) $\frac{3}{3}$ (2) $\frac{1}{3}$ (3) $1\frac{1}{3}$ (4) $1\frac{2}{3}$

2. Marta has 6 rooms to paint. She has painted $3\frac{1}{4}$ rooms. How many more rooms must she paint?

   (1) $3\frac{3}{4}$ (2) $3\frac{2}{4}$ (3) $2\frac{3}{4}$ (4) $1\frac{3}{4}$

3. The Lake Clinic reports that $\frac{5}{6}$ of its patients never come back for a second visit. How many do go back for a second visit?

   (1) $\frac{4}{6}$ (2) $\frac{11}{12}$ (3) $\frac{1}{3}$ (4) $\frac{1}{6}$

4. Michelle has 3 hours to wait for a plane. She has waited $1\frac{5}{6}$ hours. How many hours does she have left to wait?

(1) $2\frac{2}{6}$ (2) $1\frac{1}{6}$ (3) $\frac{5}{6}$ (4) $1\frac{5}{6}$

5. In the last city election, Charles Waters got $\frac{7}{12}$ of the vote. Harold Gibson got $\frac{5}{12}$ of the vote. By how much of the vote did Charles Waters win?

(1) $\frac{12}{12}$ (2) $\frac{1}{6}$ (3) $\frac{2}{24}$ (4) $\frac{12}{24}$

6. Oscar wants to take a thermos of coffee to work. He has 6 cups of coffee in his coffee maker. The thermos holds $4\frac{1}{2}$ cups. How many cups will he have left over?

(1)   1   (2) $\frac{1}{2}$ (3) $2\frac{1}{2}$ (4) $1\frac{1}{2}$

7. The Belvoir basketball team played 12 games. It won 7 of its games. What fraction of its games did it lose?

(1) $\frac{6}{12}$ (2) $\frac{4}{12}$ (3) $\frac{5}{7}$ (4) $\frac{5}{12}$

8. Mrs. Cantrell filled up with gas on the weekend. On Monday she used $\frac{3}{8}$ of a tank. On Tuesday she used $\frac{1}{8}$ of a tank. What fraction of the tank did she have left on Wednesday?

(1) $\frac{3}{8}$ (2) $\frac{3}{4}$ (3) $\frac{1}{2}$ (4) $\frac{4}{16}$

9. Superba tuna comes in $6\frac{3}{8}$-ounce cans. Tastee tuna comes in $4\frac{5}{8}$-ounce cans. How many more ounces of tuna do you get when you buy Superba?

(1) $1\frac{3}{4}$ (2) $1\frac{5}{8}$ (3) $1\frac{3}{8}$ (4) $2\frac{3}{4}$

10. Saturday Jean ran $1\frac{2}{5}$ miles. On Sunday she ran $\frac{3}{5}$ of a mile. How much farther did she run on Saturday than on Sunday?

(1) $1\frac{1}{5}$ (2) $\frac{4}{5}$ (3) 2   (4) $\frac{3}{5}$

Check your answers on page 137.

16. MORE ON ADDING AND SUBTRACTING FRACTIONS

Only the numerators (top numbers) are added when you add fractions. Only numerators are subtracted when you subtract fractions. Until now you have been adding and subtracting fractions with the same denominator (bottom number).

But how can you add these fractions?

$\frac{1}{2} + \frac{1}{4} = ?$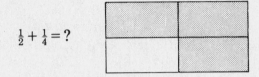

CHANGING A DENOMINATOR

You can add or subtract only if the fractions have the same denominator. So you must sometimes change a denominator to make it the same as another denominator. You can change a fraction to a different denominator by multiplying.

$$\frac{1}{2} + \frac{1}{4} = ?$$

To change the denominator 2 to 4, multiply by 2. $2 \times 2 = 4$. You must also multiply the numerator by the same number. $1 \times 2 = 2$.

$\frac{1}{2} \times \frac{2}{2} = \frac{2}{4}$ This is called <u>raising a fraction to higher terms.</u>

Notice that you multiply by $\frac{2}{2}$. $\frac{2}{2}$ equals 1. Remember that multiplying by 1 does not change the value. You are not changing the value of $\frac{1}{2}$. You are only changing its form.

Now you can add the example:

$$
\begin{array}{rcl}
 & \frac{1}{2} & = & \frac{2}{4} \\
+ & \frac{1}{4} & = & \frac{1}{4} \\
\hline
 & & & \frac{3}{4}
\end{array}
$$

Both fractions have a common denominator. Having a <u>common denominator</u> means having the same denominator.

112

Subtraction works the same way. Look at this example:

$$\frac{5}{18} - \frac{1}{6} = ?$$

You can change the denominator 6 to 18 by multiplying by 3. $3 \times 6 =$ 18. You must also multiply the numerator by the same number. $3 \times 1 = 3$. Then the problem becomes:

$$\begin{array}{r} \frac{5}{18} = \frac{5}{18} \\ - \frac{1}{6} = \frac{3}{18} \\ \hline \frac{2}{18} \end{array}$$, which can be reduced to $\frac{1}{9}$

Try It (Answers on page 138.)

Raise these fractions to higher terms.

1. $\frac{1}{4} = \frac{}{8}$ 2. $\frac{2}{5} = \frac{}{10}$ 3. $\frac{3}{4} = \frac{}{16}$ 4. $\frac{1}{2} = \frac{}{8}$ 5. $\frac{1}{2} = \frac{}{4}$

6. $\frac{5}{8} = \frac{}{24}$ 7. $\frac{1}{3} = \frac{}{9}$ 8. $\frac{2}{3} = \frac{}{9}$ 9. $\frac{3}{4} = \frac{}{8}$ 10. $\frac{3}{8} = \frac{}{16}$

Try It (Answers on page 138.)

Find a common denominator and add or subtract. Reduce your answers to lowest terms.

1. $\begin{array}{r} \frac{1}{2} \\ + \frac{2}{10} \\ \hline \end{array}$

2. $\begin{array}{r} \frac{1}{2} \\ - \frac{1}{6} \\ \hline \end{array}$

3. $\begin{array}{r} \frac{1}{3} \\ + \frac{1}{24} \\ \hline \end{array}$

4. $\begin{array}{r} \frac{1}{2} \\ + \frac{1}{6} \\ \hline \end{array}$

5. $\begin{array}{r} \frac{7}{8} \\ - \frac{1}{4} \\ \hline \end{array}$

6. $\begin{array}{r} \frac{5}{6} \\ - \frac{5}{18} \\ \hline \end{array}$

7. $\begin{array}{r} \frac{1}{4} \\ + \frac{3}{8} \\ \hline \end{array}$

8. $\begin{array}{r} \frac{7}{10} \\ - \frac{1}{5} \\ \hline \end{array}$

9. $\begin{array}{r} \frac{5}{9} \\ + \frac{17}{36} \\ \hline \end{array}$

10. $\begin{array}{r} \frac{1}{3} \\ + \frac{5}{6} \\ \hline \end{array}$

11. $\begin{array}{r} \frac{1}{4} \\ - \frac{1}{8} \\ \hline \end{array}$

12. $\begin{array}{r} \frac{19}{20} \\ - \frac{1}{4} \\ \hline \end{array}$

13. $\begin{array}{r} \frac{1}{2} \\ + \frac{1}{4} \\ \hline \end{array}$

14. $\begin{array}{r} \frac{3}{4} \\ - \frac{1}{2} \\ \hline \end{array}$

15. $\begin{array}{r} \frac{5}{8} \\ + \frac{13}{24} \\ \hline \end{array}$

FINDING THE LOWEST COMMON DENOMINATOR

Sometimes you must change more than one denominator to add or subtract. Look at this example:

$$\tfrac{1}{2} - \tfrac{1}{3} = ?$$

These fractions have different denominators. You cannot subtract them. 2 won't divide into 3 evenly. 3 won't divide into 2 evenly. Both denominators must be changed. You must find a number both 2 and 3 will divide into evenly. 6 is the smallest number both 2 and 3 will divide into evenly. There are other numbers 2 and 3 will divide into evenly, but 6 is the smallest. 6 is called the <u>lowest common denominator</u>. Multiplying the two denominators will give you a common denominator. It may not always be the lowest common denominator, but it will give you a common denominator. Here is the example worked.

$$
\begin{aligned}
\tfrac{1}{2} &= \tfrac{1}{2} \times \tfrac{3}{3} = \tfrac{3}{6}\\
-\ \tfrac{1}{3} &= \tfrac{1}{3} \times \tfrac{2}{2} = \tfrac{2}{6}\\
\hline
&\qquad\qquad\qquad\ \tfrac{1}{6}
\end{aligned}
$$

Sometimes you have to search for the lowest common denominator. You can go to multiplication tables to find the lowest common denominator. Look at this example:

$$
\begin{aligned}
&\quad\ \tfrac{5}{9}\\
+\ &\quad\ \tfrac{1}{6}
\end{aligned}
$$

9 won't divide evenly into 6. 6 won't divide evenly into 9. A quick common denominator can be found by multiplying the two denominators, $6 \times 9 = 54$. 54 is a common denominator, but it is not the lowest common denominator. To find the lowest common denominator (L.C.D.) for 9 and 6, list their multiples.

The multiples of 9 are: 9, 18, 27, 36, ...

The multiples of 6 are: 6, 12, 18, 30, 36, ...

The smallest number that 9 and 6 will divide into evenly is 18. The lowest common denominator for 9 and 6 is 18. There are other common denominators, but 18 is the lowest.

Here is the example completed.

$$
\begin{aligned}
\tfrac{5}{9} &= \tfrac{5}{9} \times \tfrac{2}{2} = \tfrac{10}{18}\\
+\ \tfrac{1}{6} &= \tfrac{1}{6} \times \tfrac{3}{3} = \tfrac{3}{18}\\
\hline
&\qquad\qquad\qquad\ \tfrac{13}{18}
\end{aligned}
$$

Try It (Answers on page 138.)

Reduce where possible.

1. $\begin{array}{r} \frac{3}{4} \\ + \ \frac{1}{3} \\ \hline \end{array}$

2. $\begin{array}{r} \frac{3}{4} \\ - \ \frac{1}{3} \\ \hline \end{array}$

3. $\begin{array}{r} \frac{2}{3} \\ + \ \frac{3}{5} \\ \hline \end{array}$

4. $\begin{array}{r} \frac{1}{3} \\ \frac{1}{2} \\ + \ \frac{1}{6} \\ \hline \end{array}$

5. $\begin{array}{r} \frac{2}{3} \\ - \ \frac{1}{2} \\ \hline \end{array}$

6. $\begin{array}{r} \frac{1}{8} \\ \frac{3}{8} \\ \frac{1}{4} \\ + \ \frac{1}{2} \\ \hline \end{array}$

7. $\begin{array}{r} \frac{1}{9} \\ + \ \frac{1}{6} \\ \hline \end{array}$

8. $\begin{array}{r} \frac{3}{8} \\ - \ \frac{1}{3} \\ \hline \end{array}$

9. $\begin{array}{r} \frac{3}{16} \\ \frac{1}{4} \\ + \ \frac{1}{8} \\ \hline \end{array}$

10. $\begin{array}{r} \frac{2}{5} \\ + \ \frac{1}{6} \\ \hline \end{array}$

11. $\begin{array}{r} \frac{1}{2} \\ - \ \frac{1}{5} \\ \hline \end{array}$

12. $\begin{array}{r} \frac{3}{4} \\ + \ \frac{1}{2} \\ \hline \end{array}$

13. $\begin{array}{r} \frac{1}{4} \\ + \ \frac{1}{6} \\ \hline \end{array}$

14. $\begin{array}{r} \frac{1}{6} \\ - \ \frac{1}{8} \\ \hline \end{array}$

15. $\begin{array}{r} \frac{5}{6} \\ - \ \frac{1}{3} \\ \hline \end{array}$

ADDING MIXED NUMBERS

You can add mixed numbers only if the fractions have the same denominator. Look at this example:

	Step 1	Step 2	Step 3	Steps 4 - 5
$\begin{array}{r} 4\frac{2}{3} \\ + \ 2\frac{1}{2} \\ \hline \end{array}$	$\begin{array}{r} 4\frac{2}{3} = 4\frac{4}{6} \\ + \ 2\frac{1}{2} = 2\frac{3}{6} \\ \hline \end{array}$	$\begin{array}{r} 4\frac{4}{6} \\ + \ 2\frac{3}{6} \\ \hline \frac{7}{6} \end{array}$	$\begin{array}{r} 4\frac{4}{6} \\ + \ 2\frac{3}{6} \\ \hline 6\frac{7}{6} \end{array}$	$\begin{array}{r} 4\frac{4}{6} \\ + \ 2\frac{3}{6} \\ \hline 6\frac{7}{6} = 6 + 1\frac{1}{6} = 7\frac{1}{6} \end{array}$

Step 1—The lowest common denominator for 2 and 3 is 6.

$$\frac{2}{3} \times \frac{2}{2} = \frac{4}{6} \qquad \frac{1}{2} \times \frac{3}{3} = \frac{3}{6}$$

Step 2—Add the fractions: $\frac{4}{6} + \frac{3}{6} = \frac{7}{6}$.

Step 3—Add the whole numbers: $4 + 2 = 6$.

Step 4—Change the improper fraction to a mixed number: $\frac{7}{6} = 1\frac{1}{6}$.

Step 5—Add the sum of the whole numbers (6) to the mixed number: $6 + 1\frac{1}{6} = 7\frac{1}{6}$.

Try It (Answers on page 139.)

Reduce answers where possible.

1. $3\frac{2}{3}$
   $+4\frac{4}{9}$

2. $2\frac{1}{5}$
   $+3\frac{1}{2}$

3. $\frac{4}{5}$
   $+3\frac{2}{3}$

4. $7\frac{1}{6}$
   $+2\frac{1}{3}$

5. $3\frac{5}{6}$
   $+7\frac{2}{3}$

6. $7\frac{3}{4}$
   $+3\frac{2}{3}$

7. $6\frac{1}{3}$
   $+5\frac{4}{5}$

8. $7\frac{3}{5}$
   $+2\frac{5}{10}$

9. $1\frac{1}{2}$
   $2\frac{1}{4}$
   $+3\frac{1}{2}$

10. $17\frac{5}{8}$
    $3\frac{1}{4}$
    $+4\frac{1}{2}$

11. $20\frac{5}{6}$
    $13\frac{5}{9}$
    $+8\frac{2}{3}$

12. $10\frac{10}{16}$
    $23\frac{1}{2}$
    $+5\frac{7}{8}$

SUBTRACTING MIXED NUMBERS

You can subtract mixed numbers if the fractions have the same denominators. Look at this example:

	Step 1	Step 2	Step 3	Step 4
$4\frac{1}{3}$ $-1\frac{3}{4}$	$4\frac{1}{3}=4\frac{4}{12}$ $-1\frac{3}{4}=1\frac{9}{12}$	$3\frac{4}{12}+\frac{12}{12}=3\frac{16}{12}$ $-1\frac{9}{12}\qquad=1\frac{9}{12}$	$3\frac{16}{12}$ $-1\frac{9}{12}$ $\overline{\frac{7}{12}}$	$3\frac{16}{12}$ $-1\frac{9}{12}$ $\overline{2\frac{7}{12}}$

Step 1—Find the lowest common denominator. $3 \times 4 = 12$.

$$\frac{1}{3} \times \frac{4}{4} = \frac{4}{12} \qquad \frac{3}{4} \times \frac{3}{3} = \frac{9}{12}$$

Step 2—You can't subtract $\frac{9}{12}$ from $\frac{4}{12}$. Borrow 1 as $\frac{12}{12}$ from the 4. Add it to the $\frac{4}{12}$.

Step 3—Subtract the fractions. $\frac{16}{12} - \frac{9}{12} = \frac{7}{12}$.

Step 4—Subtract the whole numbers. $3 - 1 = 2$.

Try It (Answers on page 139.)

Reduce answers where possible.

1. $8\frac{1}{8}$
   $-2\frac{1}{4}$

2. $7\frac{5}{6}$
   $-2\frac{3}{4}$

3. $9\frac{1}{2}$
   $-2\frac{3}{4}$

4. $4\frac{1}{2}$
   $-2\frac{7}{10}$

5. $\quad 3\frac{2}{9}$
   $\quad -2\frac{5}{6}$
   $\quad \overline{}$

6. $\quad 5\frac{1}{4}$
   $\quad -1\frac{5}{6}$
   $\quad \overline{}$

7. $\quad 14\frac{1}{6}$
   $\quad -10\frac{9}{10}$
   $\quad \overline{}$

8. $\quad 2\frac{1}{9}$
   $\quad -1\frac{1}{6}$
   $\quad \overline{}$

9. $\quad 73\frac{3}{4}$
   $\quad -14\frac{1}{2}$
   $\quad \overline{}$

10. $\quad 101\frac{1}{2}$
    $\quad -\ 9\frac{2}{3}$
    $\quad \overline{}$

11. $\quad 17\frac{5}{6}$
    $\quad -\ 8\frac{14}{15}$
    $\quad \overline{}$

12. $\quad 60\frac{3}{5}$
    $\quad -42\frac{14}{15}$
    $\quad \overline{}$

WORKING WITH WORD PROBLEMS

Read the Steps for Solving Word Problems on page xii again. Then work these problems. Find the correct answer. Reduce fractions to lowest terms where possible.

1. Jan spent $\frac{1}{6}$ of her budget on clothes. She spent $\frac{1}{4}$ on rent. She spent $\frac{1}{3}$ on a car. How much of her budget did she spend on these three items?

   (1) $\frac{3}{13}$ (2) $\frac{1}{13}$ (3) $\frac{2}{3}$ (4) $\frac{3}{4}$

2. Frank had a board $4\frac{1}{2}$ feet long. He cut off a piece $1\frac{3}{4}$ feet long. How much of the board was left?

   (1) $3\frac{1}{4}$ (2) $3\frac{1}{2}$ (3) $2\frac{3}{4}$ (4) $3\frac{3}{4}$

3. Katherine jogged $3\frac{3}{4}$ miles Saturday. She jogged $2\frac{7}{10}$ miles Sunday. How far did she jog that weekend?

   (1) $5\frac{5}{6}$ (2) $6\frac{9}{20}$ (3) $6\frac{1}{2}$ (4) $6\frac{5}{8}$

4. Mrs. Pachero bought $3\frac{1}{4}$ of a ton of coal from a dealer. Her neighbor, Mrs. Jackson, bought $2\frac{1}{10}$ of a ton of coal from the same dealer. How many tons of coal did the dealer sell to these neighbors?

   (1) $5\frac{26}{43}$ (2) $5\frac{4}{5}$ (3) $5\frac{1}{3}$ (4) $5\frac{7}{20}$

5. Mrs. Parker bought $2\frac{1}{2}$ yards of material to make a dress for her daughter. She also bought $2\frac{1}{4}$ yards of material to make a dress for herself. She bought another $4\frac{3}{8}$ yards to line her dress. How many yards of material did she buy altogether?

   (1) $8\frac{3}{8}$ (2) $9\frac{1}{8}$ (3) 10 (4) $6\frac{2}{3}$

6. A barrel of nails weighs $189\frac{1}{4}$ pounds. The barrel itself weighs $15\frac{7}{8}$ pounds. How much do the nails weigh?

   (1) $173\frac{3}{8}$ (2) $174\frac{3}{8}$ (3) $204\frac{1}{8}$ (4) $214\frac{1}{8}$

7. Kristin has four books to fit on a 12-inch shelf. The books measure $1\frac{1}{2}$ inches, $1\frac{1}{8}$ inches, $\frac{9}{16}$ inches, and $2\frac{1}{4}$ inches. How much space will she have left on the shelf?
   (1) $4\frac{7}{16}$ (2) $7\frac{9}{16}$ (3) $8\frac{9}{16}$ (4) $16\frac{7}{16}$

8. A fruit store has $26\frac{3}{4}$ pounds of tomatoes to sell. The following amounts were sold: $1\frac{1}{2}$ pounds, $3\frac{1}{4}$ pounds, $4\frac{1}{8}$ pounds, and $5\frac{3}{8}$ pounds. How many pounds were left after these sales?
   (1) $14\frac{1}{4}$ (2) $13\frac{5}{8}$ (3) 13   (4) $12\frac{1}{2}$

9. Sal's delivery truck gets $9\frac{3}{10}$ miles per gallon of gas. He has four trips to make. The first stop is $2\frac{1}{4}$ miles away. The second stop is $\frac{9}{10}$ of a mile further, the third stop is $1\frac{7}{8}$ miles further, and the fourth stop is $\frac{3}{8}$ of a mile further. How many more miles can he travel on one gallon of gas?
   (1) $3\frac{9}{10}$ (2) $3\frac{1}{5}$ (3) $4\frac{1}{10}$ (4) $14\frac{7}{10}$

10. Henry weighed 200 pounds when he started his diet. He lost $1\frac{3}{8}$ pounds the first week, $\frac{3}{4}$ pounds the second, $2\frac{5}{12}$ pounds the third, and $1\frac{1}{8}$ pounds the fourth. How much did he weigh after the fourth week?
    (1) $195\frac{1}{3}$ (2) $194\frac{1}{3}$ (3) $194\frac{1}{12}$ (4) $194\frac{17}{24}$

Check your answers on page 139.

17. MULTIPLYING FRACTIONS

Rusty can paint a wall in $\frac{1}{4}$ hour. Russell can paint the same size wall in $\frac{1}{2}$ that time. How long would it take Russell to paint the wall? To find out how long it takes Russell to paint the wall multiply: $\frac{1}{2}$ by $\frac{1}{4}$. To multiply fractions: multiply across the top, then multiply across the bottom. You don't need common denominators to multiply.

$$\frac{1}{2} \times \frac{1}{4} = \frac{1 \times 1}{2 \times 4} = \frac{1}{8}$$

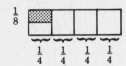

Step 1—Multiply the top numbers: $1 \times 1 = 1$.

Step 2—Multiply the bottom numbers: $2 \times 4 = 8$.

Step 3—Write the answer in the form of a fraction: $\frac{1}{8}$.

The problem $\frac{1}{2} \times \frac{1}{4}$ means the same as $\frac{1}{2}$ <u>of</u> $\frac{1}{4}$. Sometimes the answer to a multiplication problem can be reduced. Look at this example:

$$\frac{2}{3} \times \frac{1}{2} = \frac{2}{6} = \frac{2 \div 2}{6 \div 2} = \frac{1}{3}$$

Step 1—Multiply the fractions: $\frac{2}{3} \times \frac{1}{2} = \frac{2}{6}$.

Step 2—Reduce the fraction by dividing both the top and bottom numbers by 2: $\frac{2}{6} = \frac{1}{3}$.

Try It (Answers on page 140.)

Multiply. Reduce where possible.

1. $\frac{1}{2} \times \frac{3}{4} =$

2. $\frac{2}{3} \times \frac{1}{4} =$

3. $\frac{1}{3} \times \frac{3}{4} =$

4. $\frac{2}{3} \times \frac{1}{7} =$

5. $\frac{1}{3} \times \frac{6}{7} =$

6. $\frac{2}{9} \times \frac{3}{4} =$

7. $\frac{3}{8} \times \frac{3}{5} =$

8. $\frac{4}{9} \times \frac{1}{2} =$

9. $\frac{3}{8} \times \frac{2}{5} =$

10. $\frac{1}{5} \times \frac{6}{7} =$

11. $\frac{1}{2} \times \frac{1}{2} =$

12. $\frac{1}{2} \times \frac{1}{2} \times \frac{1}{2} =$

13. $\frac{1}{2} \times \frac{1}{3} =$

14. $\frac{1}{2} \times \frac{3}{2} =$

15. $\frac{1}{2} \times \frac{1}{4} \times \frac{4}{5} =$

MULTIPLYING FRACTIONS AND WHOLE NUMBERS

Carmen plans to run $\frac{3}{5}$ of a mile each day for four days. How many miles will she run altogether? To find out, you must multiply a fraction by a whole number. Give the whole number a denominator of 1. This does not change its value. Multiply and reduce if possible.

$$\frac{3}{5} \times 4 = \frac{3}{5} \times \frac{4}{1} = \frac{12}{5}, \text{ or } 2\frac{2}{5}$$

Try It (Answers on page 140.)

Multiply. Change improper fractions to mixed numbers.

1. $\frac{1}{4} \times 5 =$ 2. $\frac{1}{10} \times 3 =$ 3. $5 \times \frac{2}{3} =$

4. $4 \times \frac{2}{3} =$ 5. $\frac{7}{8} \times 7 =$ 6. $\frac{2}{3} \times 5 =$

7. $\frac{1}{6} \times 7 =$ 8. $3 \times \frac{4}{5} =$ 9. $8 \times \frac{2}{3} =$

10. $6 \times \frac{3}{5} =$ 11. $\frac{2}{9} \times 4 =$ 12. $7 \times \frac{2}{3} =$

13. $\frac{2}{7} \times 6 =$ 14. $\frac{3}{8} \times 3 =$ 15. $4 \times \frac{3}{5} =$

CANCELING

You can reduce fractions before you multiply. This is called <u>canceling</u>. When you cancel, you divide both a top number and a bottom number by the same number. It doesn't matter if the numerator and denominator are not in the same fraction. Look at this example:

	Step 1	Step 2
$\frac{1}{4} \times \frac{8}{9} =$	$\frac{1}{\cancel{4}_1} \times \frac{\cancel{8}^2}{9} =$	$\frac{1}{1} \times \frac{2}{9} = \frac{2}{9}$

Step 1—The denominator 4 and the numerator 8 can both be divided by 4. Cross out, or cancel, the 4 and the 8. Write the answers to the divisions: $4 \div 4 = 1$ and $8 \div 4 = 2$.

Step 2—Multiply the numbers in the canceled fractions: $1 \times 2 = 2$ and $1 \times 9 = 9$.

Canceling can save you the work of reducing at the end of the problem. If you do not cancel, you will have to reduce your answer.

Without canceling:

$$\frac{5}{6} \times \frac{3}{4} = \frac{15}{24} = \frac{5}{8}$$

With canceling:

$$\frac{5}{\underset{2}{\cancel{6}}} \times \frac{\overset{1}{\cancel{3}}}{4} = \frac{5}{8}$$

You can cancel more than once. Look at this example:

$$\frac{5}{12} \times \frac{4}{15} = \frac{\overset{1}{\cancel{5}}}{\underset{3}{\cancel{12}}} \times \frac{\overset{1}{\cancel{4}}}{\underset{3}{\cancel{15}}} = \frac{1}{9}$$

Each time you cancel a numerator you must cancel a denominator. In this example, the 5 and 15 were divided by 5. The 4 and 12 were divided by 4.

Try It (Answers on page 141.)

Multiply. Remember to cancel.

1. $\frac{1}{4} \times \frac{8}{9} =$

2. $\frac{9}{10} \times \frac{1}{3} =$

3. $\frac{3}{5} \times \frac{4}{15} =$

4. $\frac{6}{7} \times \frac{2}{3} =$

5. $\frac{1}{6} \times \frac{2}{3} =$

6. $\frac{2}{3} \times \frac{3}{4} =$

7. $\frac{4}{5} \times \frac{10}{13} =$

8. $\frac{3}{10} \times \frac{7}{9} =$

9. $\frac{1}{2} \times \frac{2}{3} \times \frac{3}{5} =$

10. $\frac{3}{8} \times \frac{4}{7} =$

11. $\frac{9}{10} \times \frac{5}{6} =$

12. $\frac{2}{5} \times \frac{3}{4} \times \frac{5}{6} =$

13. $\frac{8}{11} \times \frac{15}{16} =$

14. $\frac{3}{2} \times \frac{2}{3} =$

15. $\frac{4}{5} \times \frac{5}{4} =$

MULTIPLYING MIXED NUMBERS

You can multiply mixed numbers. Change them to improper fractions, then multiply. Look at this example:

	Step 1	Step 2	Step 3
$\frac{2}{3} \times 2\frac{1}{5} =$	$\frac{2}{3} \times \frac{11}{5} =$	$\frac{2}{3} \times \frac{11}{5} = \frac{22}{15}$	$\frac{22}{15} = 1\frac{7}{15}$

Step 1—Change the mixed number to an improper fraction: $2\frac{1}{5} = \frac{11}{5}$

Step 2—Multiply: $\frac{2}{3} \times \frac{11}{5} = \frac{22}{15}$.

Step 3—Change the improper fraction to a mixed number: $\frac{22}{15} = 1\frac{7}{15}$.

Try It (Answers on page 141.)

Multiply. Cancel where you can.

1. $\frac{3}{4} \times 5\frac{1}{2} =$ 2. $\frac{5}{7} \times 2\frac{2}{3} =$ 3. $1\frac{1}{2} \times 2\frac{3}{4} =$

4. $\frac{4}{5} \times 2\frac{1}{3} =$ 5. $3\frac{1}{3} \times \frac{4}{9} =$ 6. $4\frac{1}{2} \times \frac{2}{3} =$

7. $3\frac{2}{3} \times \frac{5}{6} =$ 8. $\frac{3}{5} \times 2\frac{1}{3} =$ 9. $\frac{4}{5} \times 5\frac{1}{4} =$

10. $\frac{1}{3} \times 3\frac{1}{4} =$ 11. $1\frac{3}{5} \times 4\frac{1}{2} =$ 12. $2\frac{1}{4} \times 1\frac{1}{3} =$

13. $2\frac{1}{2} \times \frac{1}{2} =$ 14. $2\frac{1}{2} \times 3\frac{3}{4} =$ 15. $1\frac{1}{2} \times 1\frac{1}{2} =$

WORKING WITH WORD PROBLEMS

Read the Steps for Solving Word Problems on page xii again. Then work these problems. Find the correct answer. Reduce fractions to lowest terms where possible.

1. Jose worked in his garden for $2\frac{1}{4}$ hours every day for 7 days. What is the total number of hours he worked?
   (1) $12\frac{1}{2}$ (2) $15\frac{3}{4}$ (3) $15\frac{1}{2}$ (4) $12\frac{3}{4}$

2. Mrs. Gross planned to work in her yard $\frac{1}{2}$ day on Saturday. She worked $\frac{2}{3}$ of her planned time. What part of the day did she work?
   (1) $\frac{1}{2}$ (2) $\frac{5}{6}$ (3) $\frac{1}{3}$ (4) $\frac{1}{4}$

3. Joan's favorite cake recipe calls for $1\frac{1}{4}$ cups of flour. She wants to make $\frac{1}{2}$ of the usual recipe. How many cups of flour does she need?
   (1) $\frac{5}{8}$ (2) $\frac{3}{4}$ (3) $\frac{3}{8}$ (4) $\frac{1}{4}$

4. Claudia bought $6\frac{2}{3}$ yards of lumber. She used $\frac{2}{3}$ of this lumber to build a dog house. How many yards of lumber does she have left?
   (1) $6\frac{1}{2}$ (2) $4\frac{4}{9}$ (3) $2\frac{2}{9}$ (4) $5\frac{1}{3}$

5. Jean lost her job last month. She could pay only $\frac{2}{5}$ of her rent this month. Her rent is 350 dollars. How much did she pay?
   (1) $70   (2) $140   (3) $150   (4) $175

To find the area of a rectangle (the amount of space inside a surface with four sides), you multiply the length and the width. The answer you get will be square units, such as square inches, square feet, or square yards. Use this information to answer questions 6, 7, and 8.

6. Delia wants to wallpaper the part of the kitchen wall behind her stove. The section of wall measures $1\frac{1}{2}$ yards by $1\frac{1}{3}$ yards. How many square yards of wallpaper does she need to cover this area?

    (1) $\frac{1}{6}$ (2) $1\frac{1}{6}$ (3) $1\frac{1}{3}$ (4) 2

7. Fred wants to buy carpet for a hallway. It measures $5\frac{1}{2}$ yards by $2\frac{1}{4}$ yards. How many square yards of carpet will he need?

    (1) $12\frac{3}{8}$ (2) $11\frac{3}{8}$ (3) 10 (4) $3\frac{1}{3}$

8. Gloria is making leather bookmarks to sell at a street fair. Each one measures $1\frac{5}{8}$ inches by 6 inches. How many square inches of leather does she need for each bookmark?

    (1) $\frac{13}{48}$ (2) $9\frac{3}{4}$ (3) $9\frac{3}{8}$ (4) $7\frac{5}{8}$

9. Randall runs a low hurdle course in $1\frac{7}{10}$ minutes. His coach wants him to run it in $\frac{5}{8}$ of that time. How fast does his coach want him to run the course?

    (1) $2\frac{1}{8}$ (2) $1\frac{5}{8}$ (3) $1\frac{7}{16}$ (4) $1\frac{1}{16}$

10. Dennis claims he can run a mile in $\frac{4}{5}$ of the time it takes Gary. Sandy says she can run a mile in $\frac{1}{2}$ of Dennis's time. Carl says he can run a mile in $\frac{3}{4}$ of Sandy's time. What fraction of Gary's time does it take Carl to run a mile?

    (1) $\frac{3}{5}$ (2) $\frac{3}{8}$ (3) $\frac{3}{10}$ (4) $2\frac{1}{20}$

Check your answers on page 141.

18. DIVIDING FRACTIONS

Dividing a number by 2 gives the same answer as multiplying by $\frac{1}{2}$.
Look at this example:

$$6 \div 2 = 3 \qquad 6 \times \tfrac{1}{2} = 3$$

$$\tfrac{6}{1} \div \tfrac{2}{1} = \tfrac{6}{1} \times \tfrac{1}{2}$$

The answers are the same. The $\frac{2}{1}$ is turned upside down and multiplied. Turning a fraction upside down is called <u>inverting.</u>

$$\tfrac{2}{1} \text{ inverted is } \tfrac{1}{2}$$

$$\tfrac{3}{5} \text{ inverted is } \tfrac{5}{3}$$

$$\tfrac{3}{4} \text{ inverted is } \tfrac{4}{3}$$

To divide by a fraction, invert the fraction you are dividing by. Then multiply.

You have a piece of wood $\frac{1}{2}$ yard long. You want to cut it into pieces $\frac{1}{6}$ yard long. How many pieces will you get? Divide $\frac{1}{2}$ by $\frac{1}{6}$.

	Step 1	Step 2
$\frac{1}{2} \div \frac{1}{6} =$	$\frac{1}{6}$ inverted is $\frac{6}{1}$	$\frac{1}{2} \times \frac{6}{1} = \frac{6}{2} = 3$

Step 1—Invert the fraction you are dividing by: $\frac{1}{6}$ inverted is $\frac{6}{1}$.

Step 2—Multiply: $\frac{1}{2} \times \frac{6}{1} = \frac{6}{2}$. Reduce your answer.

You will get three $\frac{1}{6}$-yard pieces from the $\frac{1}{2}$-yard piece of wood.

Try It (Answers on page 142.)

Reduce answers where possible.

1. $\frac{1}{2} \div \frac{1}{4} =$ 2. $\frac{1}{6} \div \frac{1}{7} =$ 3. $\frac{2}{3} \div \frac{1}{2} =$

4. $\frac{3}{4} \div \frac{2}{3} =$ 5. $\frac{4}{7} \div \frac{2}{5} =$ 6. $\frac{1}{4} \div \frac{3}{8} =$

7. $\frac{4}{5} \div \frac{1}{2} =$ 8. $\frac{5}{8} \div \frac{2}{7} =$ 9. $\frac{2}{5} \div \frac{2}{3} =$

10. $\frac{3}{8} \div \frac{2}{5} =$ 11. $\frac{4}{9} \div \frac{1}{4} =$ 12. $\frac{2}{3} \div \frac{2}{5} =$

13. $\frac{5}{6} \div \frac{6}{7} =$ 14. $\frac{5}{12} \div \frac{3}{7} =$ 15. $\frac{1}{2} \div \frac{1}{2} =$

DIVIDING WITH FRACTIONS AND WHOLE NUMBERS

You can divide a fraction by a whole number. First, change the whole number to a fraction. To make the fraction, use a denominator of 1.

$$5 = \tfrac{5}{1} \qquad 6 = \tfrac{6}{1} \qquad 7 = \tfrac{7}{1}$$

Suppose you want to divide $\tfrac{1}{2}$ pound of nuts among four people:

	Step 1	Step 2	Step 3
$\tfrac{1}{2} \div 4 =$	$\tfrac{1}{2} \div \tfrac{4}{1} =$	$\tfrac{4}{1}$ inverted is $\tfrac{1}{4}$	$\tfrac{1}{2} \times \tfrac{1}{4} = \tfrac{1}{8}$

Step 1—Change the whole number to a fraction: $4 = \tfrac{4}{1}$.

Step 2—Invert the number you are dividing by: $\tfrac{4}{1}$ inverted is $\tfrac{1}{4}$.

Step 3—Multiply: $\tfrac{1}{2} \times \tfrac{1}{4} = \tfrac{1}{8}$.

Each person will get $\tfrac{1}{8}$ pound of nuts.

Sometimes the whole is not the number you are dividing by. In this case, you still give the whole number a denominator of 1. Suppose you want to divide 4 pounds of hamburger into $\tfrac{1}{2}$-pound servings. How many servings will you have?

	Step 1	Step 2	Step 3
$4 \div \tfrac{1}{2} =$	$\tfrac{4}{1} \div \tfrac{1}{2} =$	$\tfrac{1}{2}$ inverted is $\tfrac{2}{1}$	$\tfrac{4}{1} \times \tfrac{2}{1} = \tfrac{8}{1}$, or 8

Step 1—Change the whole number to a fraction: $4 = \tfrac{4}{1}$.

Step 2—Invert the number you are dividing by: $\tfrac{1}{2}$ inverted is $\tfrac{2}{1}$.

Step 3—Multiply: $\tfrac{4}{1} \times \tfrac{2}{1} = \tfrac{8}{1}$ or 8.

You will have eight $\tfrac{1}{2}$-pound servings.

Try It (Answers on page 142.)

Reduce answers where possible.

1. $\tfrac{2}{3} \div 3 =$ 2. $\tfrac{4}{5} \div 3 =$ 3. $2 \div \tfrac{1}{2} =$

4. $\tfrac{1}{6} \div 2 =$ 5. $\tfrac{7}{8} \div 2 =$ 6. $\tfrac{1}{2} \div 2 =$

7. $4 \div \tfrac{3}{5} =$ 8. $\tfrac{2}{3} \div 5 =$ 9. $4 \div \tfrac{2}{3} =$

10. $3 \div \frac{5}{8} =$ 11. $6 \div \frac{1}{5} =$ 12. $\frac{1}{5} \div 5 =$

13. $\frac{3}{4} \div 4 =$ 14. $\frac{3}{7} \div 4 =$ 15. $5 \div \frac{1}{5} =$

CANCELING IN DIVISION

Sometimes you can cancel in division. <u>Invert before you cancel, and multiply after you cancel.</u> Look at this example:

	Step 1	Step 2	Step 3
$\frac{2}{3} \div \frac{1}{6} =$	$\frac{2}{3} \times \frac{6}{1} =$	$\frac{2}{\underset{1}{3}} \times \frac{\overset{2}{6}}{1} =$	$\frac{2}{1} \times \frac{2}{1} = \frac{4}{1}$, or 4

Step 1—Invert and change to multiplication: $\frac{2}{3} \div \frac{1}{6} = \frac{2}{3} \times \frac{6}{1}$.

Step 2—You can divide the 3 and the 6 by 3. Cancel: $\frac{2}{\underset{1}{3}} \times \frac{\overset{2}{6}}{1} = \frac{2}{1} \times \frac{2}{1}$.

Step 3—Multiply: $\frac{2}{1} \times \frac{2}{1} = \frac{4}{1}$ or 4.

Try It (Answers on page 142.)

Divide. Remember to invert first, then cancel. Reduce where possible.

1. $\frac{1}{4} \div \frac{1}{8} =$ 2. $\frac{5}{6} \div \frac{5}{12} =$ 3. $\frac{2}{3} \div \frac{1}{3} =$

4. $\frac{3}{5} \div \frac{6}{7} =$ 5. $\frac{4}{5} \div \frac{3}{5} =$ 6. $\frac{1}{5} \div \frac{3}{10} =$

7. $\frac{4}{7} \div \frac{8}{9} =$ 8. $\frac{3}{8} \div \frac{1}{6} =$ 9. $\frac{1}{2} \div \frac{1}{4} =$

10. $\frac{3}{5} \div \frac{9}{10} =$ 11. $\frac{2}{3} \div \frac{1}{9} =$ 12. $\frac{3}{8} \div \frac{3}{4} =$

13. $\frac{1}{6} \div \frac{1}{6} =$ 14. $\frac{2}{7} \div \frac{4}{7} =$ 15. $\frac{3}{4} \div \frac{1}{2} =$

DIVIDING AND MIXED NUMBERS

Before dividing, change <u>any</u> mixed numbers to improper fractions. Look at this example.

	Step 1	Step 2	Step 3
$1\frac{1}{4} \div \frac{2}{3} =$	$\frac{5}{4} \div \frac{2}{3} =$	$\frac{5}{4} \times \frac{3}{2} = \frac{15}{8}$	$\frac{15}{8} = 1\frac{7}{8}$

Step 1—Change the mixed number to an improper fraction: $1\frac{1}{4} = \frac{5}{4}$

Step 2—Invert and multiply: $\frac{5}{4} \times \frac{3}{2} = \frac{15}{8}$.

Step 3—Reduce your answer. Change improper fractions back to mixed numbers: $\frac{15}{8} = 1\frac{7}{8}$.

Try It (Answers on page 143.)

1. $1\frac{3}{4} \div \frac{1}{4} =$ 2. $3\frac{1}{2} \div \frac{3}{4} =$ 3. $1\frac{1}{2} \div 1\frac{1}{2} =$

4. $8 \div 1\frac{2}{3} =$ 5. $2\frac{1}{4} \div \frac{1}{4} =$ 6. $3\frac{1}{2} \div 2\frac{1}{4} =$

7. $2\frac{1}{2} \div \frac{3}{4} =$ 8. $5\frac{1}{2} \div 3 =$ 9. $4 \div 1\frac{1}{3} =$

10. $1\frac{1}{2} \div \frac{1}{5} =$ 11. $3\frac{1}{8} \div 2\frac{1}{2} =$ 12. $2\frac{2}{3} \div 1\frac{1}{2} =$

13. $\frac{2}{5} \div 1\frac{1}{7} =$ 14. $3\frac{3}{4} \div 3\frac{3}{4} =$ 15. $8\frac{1}{3} \div 1\frac{2}{3} =$

WORKING WITH WORD PROBLEMS

Read the Steps for Solving Word Problems on page xii again. Then work these problems. Find the correct answer. Reduce fractions to lowest terms where possible.

1. Linda is having a wedding shower for her friend. She bought 3 pies and cut each of them into eighths. How many pieces of pie did she have altogether?
   (1) 12   (2) $1\frac{3}{8}$ (3) $\frac{1}{24}$ (4) 24

2. Jacob cut a board $1\frac{1}{4}$ yards long into 5 equal pieces. How many yards long was each piece?
   (1) $\frac{1}{5}$ (2) $\frac{1}{4}$ (3) $\frac{2}{5}$ (4) $\frac{3}{4}$

3. The Touring Club bicycled 18 miles in $4\frac{1}{2}$ hours. How many miles did they bicycle in 1 hour?
   (1) 6   (2) $4\frac{1}{2}$ (3) $2\frac{1}{8}$ (4) 4

4. Maria walked a total $3\frac{3}{8}$ miles to work in 9 days. How many miles did she walk in 3 days?
   (1) $1\frac{1}{8}$ (2) $2\frac{1}{4}$ (3) $\frac{3}{8}$ (4) $\frac{1}{4}$

5. Mrs. Garcia made $6\frac{1}{2}$ gallons of iced tea. She wants to keep the tea in $1\frac{1}{2}$ gallon jars. How many jars does she need?
   (1) 4   (2) 5   (3) 6   (4) 3

6. Marie has $\frac{5}{8}$ feet of ribbon. She wants to cut it into $\frac{1}{4}$-foot pieces. How many pieces will she get?
   (1) $\frac{5}{24}$ (2) $2\frac{1}{8}$ (3) $2\frac{1}{2}$ (4) $3\frac{1}{8}$

7. Bill bought $2\frac{1}{3}$ pounds of ground chuck to make hamburgers. He wants each burger to weigh about 7 ounces. (One ounce equals $\frac{1}{16}$ of a pound.) How many burgers will he get?
   (1) 37　　(2) 5　　(3) 1　　(4) $\frac{1}{3}$

8. A work crew of 14 people are assigned to clear a $\frac{7}{8}$-acre field. The work is to be divided equally. How much of the field will each person clear?
   (1) 16　　(2) $\frac{1}{112}$　　(3) $\frac{7}{16}$　　(4) $\frac{1}{16}$

9. Kathy bought $1\frac{7}{8}$ pounds of nuts to make cookies. She wants to make five batches of cookies. How many pounds of nuts does she have for each batch?
   (1) $\frac{3}{8}$　　(2) $\frac{17}{40}$　　(3) $\frac{14}{40}$　　(4) $\frac{3}{40}$

10. On her diet, Freda lost $5\frac{2}{3}$ pounds. Her doctor's charts show she lost $1\frac{8}{9}$ pounds a week. How many weeks did it take her to lose the $5\frac{2}{3}$ pounds?
    (1) $7\frac{5}{9}$　　(2) 4　　(3) $3\frac{7}{9}$　　(4) 3

Check your answers on page 143.

19. MEASURING YOUR PROGRESS

TEST A

Add or Subtract

1. $\begin{array}{r} 1 \\ -\ \ \frac{5}{8} \\ \hline \end{array}$

2. $\frac{2}{5} - \frac{1}{5} =$

3. $\begin{array}{r} \frac{7}{8} \\ -\ \frac{3}{8} \\ \hline \end{array}$

4. $\begin{array}{r} 25 \\ -\ \ \frac{9}{12} \\ \hline \end{array}$

5. $\begin{array}{r} \frac{3}{16} \\ \frac{3}{4} \\ +\ \frac{1}{2} \\ \hline \end{array}$

6. $\begin{array}{r} \frac{7}{8} \\ -\ \frac{1}{3} \\ \hline \end{array}$

7. $\begin{array}{r} 16\frac{1}{8} \\ -\ \ \frac{5}{8} \\ \hline \end{array}$

8. $\begin{array}{r} 10\frac{1}{6} \\ 7\frac{1}{4} \\ +\ 3\frac{7}{12} \\ \hline \end{array}$

9. $\begin{array}{r} 8\frac{3}{5} \\ -\ 3\frac{2}{3} \\ \hline \end{array}$

10. $\frac{5}{9} + \frac{3}{9} + \frac{1}{9} =$

11. $\begin{array}{r} 36\frac{8}{14} \\ +\ 31\frac{6}{14} \\ \hline \end{array}$

12. $\begin{array}{r} 21\frac{2}{3} \\ +\ 36\frac{2}{12} \\ \hline \end{array}$

13. $\begin{array}{r} 9\frac{2}{3} \\ -\ 6\frac{11}{12} \\ \hline \end{array}$

14. $\begin{array}{r} 4\frac{7}{9} \\ 12\frac{5}{6} \\ +\ 1\frac{1}{4} \\ \hline \end{array}$

15. $\begin{array}{r} 18\frac{9}{10} \\ +\ \ \frac{1}{4} \\ \hline \end{array}$

16. $\begin{array}{r} 5 \\ -\ 2\frac{5}{8} \\ \hline \end{array}$

17. $\begin{array}{r} 45\frac{11}{15} \\ -\ 27\frac{8}{15} \\ \hline \end{array}$

18. $\begin{array}{r} 13\frac{2}{3} \\ -\ \ 8\frac{1}{4} \\ \hline \end{array}$

19. $\begin{array}{r} 210\frac{5}{9} \\ -\ 110\frac{1}{3} \\ \hline \end{array}$

20. $\begin{array}{r} 15\frac{3}{11} \\ -\ \ 9\frac{8}{11} \\ \hline \end{array}$

Multiply or Divide

21. $\frac{3}{4} \div \frac{6}{8} =$

22. $\frac{1}{3} \times 6 =$

23. $\frac{3}{4} \div 2 =$

24. $3 \div \frac{8}{9} =$

25. $\frac{3}{4} \times 2\frac{2}{3} =$

26. $\frac{1}{3} \times \frac{9}{18} =$

27. $11\frac{1}{3} \div 2\frac{5}{6} =$

28. $2\frac{3}{16} \div 1\frac{1}{4} =$

29. $1\frac{1}{4} \times 1\frac{3}{5} =$

30. $3\frac{3}{4} \times 1\frac{1}{5} =$

31. $1\frac{1}{4} \div \frac{1}{2} =$

32. $2\frac{1}{3} \div 2\frac{1}{3} =$

33. $\frac{2}{3} \times \frac{3}{5} \times \frac{1}{2} =$

34. $\frac{1}{6} \div \frac{11}{16} =$

35. $\frac{1}{8} \times 32 =$

36. $\frac{5}{6} \div 10 =$

37. $\frac{2}{3} \times 5\frac{2}{5} =$

38. $6\frac{1}{8} \div 4\frac{1}{4} =$

39. $15 \div 3\frac{1}{8} =$

40. $7\frac{3}{4} \times 15 =$

TEST B

Read the Steps for Solving Word Problems on page xii again. Look for clue words in these problems. The clue words will help you decide whether to add, subtract, multiply or divide.

1. Barbara is going for an important interview. She made a new 2-piece dress. It took $1\frac{1}{4}$ yards for one part and $2\frac{2}{4}$ yards for the other. How much material did she use altogether?

    (1) $3\frac{3}{4}$ (2) 4   (3) $2\frac{3}{4}$ (4) $3\frac{1}{4}$

2. An express train from New York to Washington, D.C., takes $3\frac{1}{3}$ hours. Regular trains take $4\frac{2}{3}$ hours to make the same trip. How many hours faster is the express?

    (1) $2\frac{1}{3}$ (2) $\frac{2}{3}$ (3) $1\frac{1}{3}$ (4) $3\frac{2}{3}$

3. Mrs. Sandrovitch made 50 cookies last Saturday. Her daughter, Susan, ate $\frac{1}{5}$ of the cookies. How many cookies did she eat?

    (1) 10   (2) 25   (3) 20   (4) 30

4. Mr. Jackson bought 6 logs to make a fence in his backyard. He sawed each log into thirds. How many pieces does he have?

    (1) 2   (2) 6   (3) 18   (4) 24

5. Ernie is baking two cakes for a party. He needs $2\frac{3}{5}$ cups of sifted flour to make a plain cake and $3\frac{1}{5}$ cups for a coconut cake. Find the total amount of flour he needs for both cakes.

    (1)   5   (2) $5\frac{4}{5}$ (3) $4\frac{3}{5}$ (4) $4\frac{1}{5}$

6. Rosa is trying to lose weight. Two months ago she weighed $135\frac{3}{4}$ pounds. Now she weighs $124\frac{1}{4}$ pounds. How many pounds did she lose?

    (1) $12\frac{3}{4}$ (2) $11\frac{1}{2}$ (3) $10\frac{1}{4}$ (4) $12\frac{1}{4}$

7. Carlos works after school at a restaurant. Last week he worked $4\frac{1}{8}$ hours on Monday, $2\frac{3}{8}$ hours on Wednesday, and 5 hours on Friday. How many hours did he work altogether?

    (1) $12\frac{3}{8}$ (2) $11\frac{1}{2}$ (3) $10\frac{3}{4}$ (4) $9\frac{5}{8}$

8. Anabelle bought a piece of ribbon 72 inches long. She used $\frac{3}{4}$ of the ribbon to trim a dress. How many inches did she have left?

    (1) 18   (2) 12   (3) 54   (4) 16

9. Marsha divided 2 apples into halves. How many pieces of apple did she have?

    (1)   4   (2)   1   (3)   2   (4) $1\frac{1}{2}$

10. Joe had a board 8 feet long. He wants to make shelves $2\frac{1}{3}$ feet long. How many shelves can he make from the board he has?

    (1) 2   (2) 3   (3) 4   (4) 5

11. To build a bookcase, Fran worked $2\frac{5}{6}$ hours on Friday. She worked $3\frac{1}{4}$ hours on Saturday and $4\frac{2}{3}$ hours on Sunday. How many hours did she work altogether?

    (1) $9\frac{3}{4}$ (2) $10\frac{3}{4}$ (3) $10\frac{2}{3}$ (4) $10\frac{11}{12}$

12. Ben bought $\frac{5}{8}$ of a pound of putty to repair a broken window. He only used $\frac{1}{3}$ of it. How much does he have left?

    (1) $\frac{7}{24}$ (2) $\frac{5}{24}$ (3) $\frac{15}{16}$ (4) $\frac{5}{12}$

13. A wooden crate full of lettuce weighs $6\frac{1}{3}$ pounds. An empty crate weighs $1\frac{5}{6}$ pounds. How much does the lettuce weigh?

    (1) $4\frac{1}{2}$ (2) $4\frac{1}{3}$ (3) $5\frac{1}{2}$ (4) $8\frac{1}{6}$

14. The scale on a road map is $\frac{1}{16}$ of an inch equals one mile. The distance between two cities measures $\frac{3}{4}$ of an inch. How many miles apart are the two cities?

    (1) $\frac{3}{64}$ (2) $\frac{1}{12}$ (3) 12   (4) 48

15. Lorna makes $11\frac{3}{10}$ dollars an hour ($\frac{3}{10}$ of a dollar is another way of saying 30¢) as a painter. Her union wants to raise her rate of pay $1\frac{1}{3}$ times higher. How much will she make if she gets the raise?

    (1) $12\frac{3}{5}$ (2) $12\frac{19}{30}$ (3) $15\frac{1}{15}$ (4) $15\frac{1}{5}$

16. The Highland Community Center bought three kinds of material to make quilts. One piece measures $4\frac{5}{6}$ yards, the second measures $3\frac{2}{3}$ yards, and the third measures $2\frac{5}{9}$ yards. How many yards of material do they have altogether?

    (1) $9\frac{1}{18}$ (2) $11\frac{1}{18}$ (3) $10\frac{1}{36}$ (4) $11\frac{1}{8}$

17. Ralph bought $3\frac{1}{3}$ pounds of ground round to make hamburgers. He wants each burger to weigh $\frac{5}{8}$ of a pound (10 ounces). How many burgers can he make?

    (1)   1   (2)   2   (3)   4   (4)   5

18. Louise is making drapes from material that is 24 inches wide. She folds the material under $\frac{5}{16}$ of an inch on both edges for the seam between sections. How wide will each section be?
(1) $24\frac{5}{8}$ (2) $23\frac{3}{8}$ (3) $23\frac{11}{16}$ (4) $24\frac{3}{8}$

19. Central City set aside $\frac{2}{3}$ of an acre to be used as a public garden. Eight people show up to plant flowers. How much land will each person get?
(1) $\frac{1}{12}$ (2) $\frac{1}{24}$ (3) $\frac{2}{12}$ (4) $\frac{5}{24}$

20. Rusty's foreman told him that he should average $3\frac{1}{7}$ jobs an hour. If Rusty works 35 hours a week, how many jobs should he do per week?
(1) 770   (2) 110   (3) 105   (4) $38\frac{1}{7}$

21. Claire's punchbowl holds $5\frac{1}{2}$ gallons. Jan saw her pour eight $\frac{1}{5}$-gallon bottles of ginger ale into the punch. How much of the punch is not ginger ale?
(1) $4\frac{9}{10}$ (2) 4   (3) $3\frac{9}{10}$ (4) $2\frac{7}{10}$

22. Mark lost $26\frac{2}{3}$ pounds on his diet. He says he lost an average of $6\frac{2}{3}$ pounds a week. How many weeks did it take him to lose $26\frac{2}{3}$ pounds?
(1) 40   (2) $33\frac{1}{3}$ (3) 20   (4) 4

23. Beth bought a $2\frac{1}{2}$-pound roast for dinner. About $\frac{1}{4}$ of the roast was fat. What did the meat in the roast weigh?
(1) $\frac{5}{8}$ (2) $1\frac{1}{2}$ (3) $1\frac{7}{8}$ (4) 2

24. Kathy cooked a $10\frac{1}{2}$-pound ham for a holiday dinner. If 14 people will be at dinner, how much meat will each person get?
(1) $\frac{1}{2}$ (2) $\frac{5}{7}$ (3) $\frac{3}{4}$ (4) $1\frac{1}{3}$

Check your answers on page 144.

ANSWERS

HOW MUCH DO YOU KNOW ALREADY?, PAGE 94

Put an X by each problem you missed. Put an X by each problem you did not do. The problems are listed under the titles of the chapters in this book. If you have <u>one or more</u> Xs in a section, study that chapter to improve your skills.

ADDING FRACTIONS (pages 102–106)

1. $\frac{3}{8}$　2. $\frac{9}{11}$　3. **1**　4. $5\frac{5}{6}$　5. **8**　6. $6\frac{2}{5}$

SUBTRACTING FRACTIONS (pages 107–111)

7. $\frac{3}{8}$　8. $\frac{1}{3}$　9. $2\frac{1}{4}$　10. $4\frac{5}{9}$　11. $4\frac{2}{3}$　12. $5\frac{2}{3}$

MORE ON ADDING AND SUBTRACTING FRACTIONS (pages 112–118)

13. $\frac{7}{8}$　14. $7\frac{1}{4}$　15. $\frac{1}{6}$　16. $\frac{1}{4}$　17. $1\frac{4}{15}$　18. $\frac{7}{12}$

MULTIPLYING FRACTIONS (pages 119–123)

19. $\frac{1}{9}$　20. $\frac{1}{2}$　21. $1\frac{7}{8}$　22. $2\frac{1}{6}$　23. $\frac{4}{7}$　24. $9\frac{1}{3}$

DIVIDING FRACTIONS (pages 124–128)

25. **1**　26. **4**　27. $\frac{4}{15}$　28. **5**　29. **2**　30. $\frac{16}{21}$

TRY IT, PAGE 95

1. $\frac{1}{4},\frac{3}{4}$　2. $\frac{3}{8},\frac{5}{8}$　3. $\frac{5}{6},\frac{1}{6}$　4. $\frac{2}{5},\frac{3}{5}$　5. $\frac{1}{3},\frac{2}{3}$　6. $\frac{1}{8},\frac{7}{8}$

TRY IT, PAGE 96

1. $\frac{1}{2},\frac{1}{2}$　2. $\frac{3}{8},\frac{5}{8}$　3. $\frac{1}{4},\frac{3}{4}$　4. $\frac{7}{10},\frac{3}{10}$　5. $\frac{5}{6},\frac{1}{6}$　6. $\frac{3}{4},\frac{1}{4}$

TRY IT, PAGE 97

1. $\frac{4}{3}$　2. $\frac{23}{17}$　3. $\frac{9}{5}$　4. $\frac{18}{15}$　5. $\frac{6}{3}$

6. $\frac{7}{3}$　7. $\frac{6}{2}$　8. $\frac{20}{13}$　9. $\frac{13}{8}$　10. $\frac{3}{2}$

TRY IT, PAGE 98, BOTTOM

1. $2\frac{1}{2}$ 2. $2\frac{1}{9}$ 3. $1\frac{2}{7}$ 4. $1\frac{1}{3}$

5. $1\frac{3}{4}$ 6. $4\frac{1}{6}$ 7. $2\frac{2}{5}$ 8. $3\frac{1}{2}$

9. $2\frac{1}{7}$ 10. $3\frac{3}{10}$ 11. $4\frac{2}{3}$ 12. $1\frac{3}{7}$

13. $4\frac{1}{2}$ 14. $1\frac{5}{8}$ 15. $1\frac{7}{8}$ 16. $4\frac{4}{5}$

17. $2\frac{1}{5}$ 18. $1\frac{9}{11}$ 19. $2\frac{1}{10}$ 20. $2\frac{3}{8}$

TRY IT, PAGE 99

1. $\frac{5}{2}$ 2. $\frac{34}{5}$ 3. $\frac{10}{3}$

4. $\frac{17}{4}$ 5. $\frac{25}{2}$ 6. $\frac{3}{2}$

7. $\frac{22}{4}$ 8. $\frac{47}{3}$ 9. $\frac{15}{4}$

10. $\frac{25}{8}$ 11. $\frac{37}{9}$ 12. $\frac{29}{5}$

13. $\frac{32}{3}$ 14. $\frac{43}{5}$ 15. $\frac{53}{5}$

TRY IT, PAGE 101

1. $\frac{1}{3}$ 2. $\frac{6}{7}$ 3. $\frac{1}{2}$ 4. $\frac{3}{7}$

5. $\frac{1}{2}$ 6. $\frac{1}{4}$ 7. $\frac{1}{2}$ 8. $\frac{2}{3}$

9. $\frac{1}{3}$ 10. $\frac{2}{5}$ 11. $\frac{1}{2}$ 12. $\frac{1}{2}$

13. $\frac{1}{3}$ 14. $\frac{2}{3}$ 15. $\frac{3}{4}$ 16. $\frac{1}{3}$

17. $\frac{3}{5}$ 18. $\frac{1}{3}$ 19. $\frac{2}{5}$ 20. $\frac{2}{3}$

TRY IT, PAGE 102, TOP

1. $\frac{3}{4}$ 2. $\frac{13}{15}$ 3. $\frac{9}{20}$

4. $\frac{9}{10}$ 5. $\frac{5}{6}$ 6. $\frac{2}{3}$

7. $\frac{4}{5}$ 8. $\frac{11}{12}$ 9. $\frac{5}{7}$

10. $\frac{6}{7}$ 11. $\frac{17}{19}$ 12. $\frac{7}{8}$

13. $\frac{7}{8}$ 14. $\frac{9}{13}$ 15. $\frac{4}{5}$

TRY IT, PAGE 102, BOTTOM

1. $\frac{4}{4} = 1$ 2. $\frac{8}{8} = 1$ 3. $\frac{2}{2} = 1$

4. $\frac{6}{7}$ 5. $\frac{4}{5}$ 6. $\frac{18}{20} = \frac{9}{10}$

7. $\frac{3}{3} = 1$ 8. $\frac{7}{9}$ 9. $\frac{3}{3} = 1$

10. $\frac{10}{12} = \frac{5}{6}$ 11. $\frac{12}{12} = 1$ 12. $\frac{3}{4}$

13. $\frac{8}{8} = 1$ 14. $\frac{9}{11}$ 15. $\frac{6}{8} = \frac{3}{4}$

TRY IT, PAGE 103

1. $7\frac{2}{3}$ 2. $11\frac{1}{3}$ 3. $10\frac{3}{5}$ 4. $13\frac{5}{9}$ 5. $40\frac{1}{2}$

6. $5\frac{1}{2}$ 7. $8\frac{6}{7}$ 8. $11\frac{2}{3}$ 9. $21\frac{1}{2}$ 10. $30\frac{6}{7}$

TRY IT, PAGE 104, TOP

1. **16**   2. **8**   3. **28**   4. **26**   5. **11**
6. **27**   7. **13**   8. **40**   9. **200**   10. **10**

TRY IT, PAGE 104, BOTTOM

1. $7\frac{1}{4}$ 2. $6\frac{2}{7}$ 3. $5\frac{1}{3}$ 4. $8\frac{1}{4}$ 5. $8\frac{2}{5}$

6. $2\frac{1}{9}$ 7. $8\frac{3}{7}$ 8. $9\frac{3}{10}$ 9. $12\frac{1}{7}$ 10. $3\frac{1}{5}$

11. $10\frac{1}{2}$ 12. $1\frac{3}{4}$ 13. $4\frac{1}{3}$ 14. $9\frac{4}{9}$ 15. 10

WORKING WITH WORD PROBLEMS, PAGE 105

1. **(4)** $\frac{1}{3}+\frac{1}{3}=\frac{2}{3}$ 2. **(2)** $\frac{1}{5}+\frac{2}{5}=\frac{3}{5}$ 3. **(3)** $\frac{1}{8}+\frac{3}{8}+\frac{1}{8}=\frac{5}{8}$

4. **(2)** $\frac{3}{20}+\frac{1}{20}+\frac{1}{20}=\frac{1}{4}$ 5. **(3)** $\frac{2}{5}+\frac{2}{5}=\frac{4}{5}$ 6. **(4)** $\frac{1}{6}+\frac{2}{6}+\frac{2}{6}=\frac{5}{6}$

7. **(4)**
$$\begin{array}{r} 2\frac{1}{3} \\ +\ 2\frac{1}{3} \\ \hline \mathbf{4\frac{2}{3}} \end{array}$$

8. **(4)** $\frac{1}{2}+\frac{1}{2}=\frac{2}{2}=\mathbf{1\ yd.}$

9. **(1)**
$$\begin{array}{r} 5\frac{1}{4} \\ +\ 2\frac{3}{4} \\ \hline 7\frac{4}{4}=\mathbf{8} \end{array}$$

10. **(3)**
$$\begin{array}{r} 1\frac{3}{4} \\ +\ 1\frac{3}{4} \\ \hline 1\frac{6}{4}=2\frac{2}{4}=\mathbf{2\frac{1}{2}} \end{array}$$

TRY IT, PAGE 107

1. $\frac{3}{5}$ 2. $\frac{3}{16}$ 3. $\frac{11}{23}$

4. $\frac{2}{5}$ 5. $\frac{2}{7}$ 6. $\frac{13}{24}$

7. $\frac{3}{10}$ 8. $\frac{8}{13}$ 9. $\frac{5}{12}$

10. $\frac{5}{12}$ 11. $\frac{1}{15}$ 12 $\frac{7}{16}$

13. $\frac{3}{8}$ 14. $\frac{1}{5}$ 15. $\frac{7}{20}$

TRY IT, PAGE 108, TOP

1. $\frac{1}{2}$ 2. $\frac{5}{8}$ 3. $\frac{7}{16}$ 4. $\frac{4}{9}$

5. $\frac{1}{3}$ 6. $\frac{6}{11}$ 7. $\frac{5}{12}$ 8. $\frac{5}{6}$

9. $\frac{7}{8}$ 10. $\frac{1}{8}$ 11. $\frac{7}{10}$ 12. $\frac{3}{5}$

TRY IT, PAGE 108, BOTTOM

1. $3\frac{1}{6}$ 2. $6\frac{2}{3}$ 3. $2\frac{1}{2}$ 4. $8\frac{1}{4}$

5. $15\frac{1}{10}$ 6. $34\frac{5}{13}$ 7. $89\frac{14}{15}$ 8. $62\frac{3}{8}$

9. $19\frac{1}{8}$ 10. $199\frac{4}{5}$ 11. $103\frac{7}{10}$ 12. $90\frac{5}{8}$

TRY IT, PAGE 109

1. $8\frac{5}{9}$ 2. $12\frac{2}{5}$ 3. $3\frac{2}{7}$ 4. $20\frac{3}{11}$

5. $1\frac{1}{3}$ 6. 1 7. $9\frac{1}{2}$ 8. $9\frac{1}{3}$

9. $4\frac{1}{2}$ 10. 129 11. $92\frac{1}{4}$ 12. $8\frac{2}{5}$

TRY IT, PAGE 110, TOP

1. $4\frac{2}{3}$ 2. $3\frac{4}{5}$ 3. $8\frac{5}{8}$ 4. $12\frac{4}{5}$

5. $11\frac{11}{12}$ 6. $6\frac{7}{9}$ 7. $83\frac{9}{11}$ 8. $111\frac{2}{3}$

9. $9\frac{5}{6}$ 10. $91\frac{3}{4}$ 11. $124\frac{3}{4}$ 12. $106\frac{3}{8}$

WORKING WITH WORD PROBLEMS, PAGE 110

1. **(3)**
$$\begin{array}{r} 1\frac{2}{3} \\ -\ \frac{1}{3} \\ \hline 1\frac{1}{3} \end{array}$$

2. **(3)**
$$\begin{array}{r} 6 = 5\frac{4}{4} \\ -\ 3\frac{1}{4} = 3\frac{1}{4} \\ \hline 2\frac{3}{4} \end{array}$$

3. **(4)**
$$\begin{array}{r} 1 = \frac{6}{6} \\ -\ \frac{5}{6} = \frac{5}{6} \\ \hline \frac{1}{6} \end{array}$$

4. **(2)**
$$\begin{array}{r} 3 = 2\frac{6}{6} \\ -\ 1\frac{5}{6} = 1\frac{5}{6} \\ \hline 1\frac{1}{6} \end{array}$$

5. **(2)**
$$\begin{array}{r} \frac{7}{12} \\ -\ \frac{5}{12} \\ \hline \frac{2}{12} = \frac{1}{6} \end{array}$$

6. **(4)**
$$\begin{array}{r} 6 = 5\frac{2}{2} \\ -\ 4\frac{1}{2} = 4\frac{1}{2} \\ \hline 1\frac{1}{2} \end{array}$$

7. **(4)**

$$\begin{array}{r} 12 \\ -\ 7 \\ \hline \end{array}$$

5, which is 5 of 12 or $\frac{5}{12}$

8. **(3)**

$$\begin{array}{r} \frac{1}{8} \\ +\ \frac{3}{8} \\ \hline \frac{4}{8} = \frac{1}{2} \end{array}$$

$$\begin{array}{r} 1 \ = \ \frac{2}{2} \\ -\ \frac{1}{2} \ = \ \frac{1}{2} \\ \hline \frac{1}{2} \end{array}$$

9. **(1)**

$$\begin{array}{r} 6\frac{3}{8} = 5\frac{11}{8} \\ -4\frac{5}{8} = 4\frac{5}{8} \\ \hline 1\frac{6}{8} = \mathbf{1\frac{3}{4}} \end{array}$$

10. **(2)**

$$\begin{array}{r} 1\frac{2}{5} = \frac{7}{5} \\ -\ \frac{3}{5} = \frac{3}{5} \\ \hline \frac{4}{5} \end{array}$$

TRY IT, PAGE 113, TOP

1. $\frac{2}{8}$ 2. $\frac{4}{10}$ 3. $\frac{12}{16}$ 4. $\frac{4}{8}$ 5. $\frac{2}{4}$

6. $\frac{15}{24}$ 7. $\frac{3}{9}$ 8. $\frac{6}{9}$ 9. $\frac{6}{8}$ 10. $\frac{6}{16}$

TRY IT, PAGE 113, BOTTOM

1. $\frac{7}{10}$ 2. $\frac{1}{3}$ 3. $\frac{3}{8}$

4. $\frac{2}{3}$ 5. $\frac{5}{8}$ 6. $\frac{5}{9}$

7. $\frac{5}{8}$ 8. $\frac{1}{2}$ 9. $\mathbf{1\frac{1}{36}}$

10. $\mathbf{1\frac{1}{6}}$ 11. $\frac{1}{8}$ 12. $\frac{7}{10}$

13. $\frac{3}{4}$ 14. $\frac{1}{4}$ 15. $\mathbf{1\frac{1}{6}}$

TRY IT, PAGE 115

1. $\mathbf{1\frac{1}{12}}$ 2. $\frac{5}{12}$ 3. $\mathbf{1\frac{4}{15}}$

4. **1**   5. $\frac{1}{6}$ 6. $\mathbf{1\frac{1}{4}}$

7. $\frac{5}{18}$ 8. $\frac{1}{24}$ 9. $\frac{9}{16}$

10. $\frac{17}{30}$ 11. $\frac{3}{10}$ 12. $\mathbf{1\frac{1}{4}}$

13. $\frac{5}{12}$ 14. $\frac{1}{24}$ 15. $\frac{1}{2}$

TRY IT, PAGE 116, TOP

1. $8\frac{1}{9}$ 2. $5\frac{7}{10}$ 3. $4\frac{7}{15}$ 4. $9\frac{1}{2}$

5. $11\frac{1}{2}$ 6. $11\frac{5}{12}$ 7. $12\frac{2}{15}$ 8. $10\frac{1}{10}$

9. $7\frac{1}{4}$ 10. $25\frac{3}{8}$ 11. $43\frac{1}{18}$ 12. 40

TRY IT, PAGE 116, BOTTOM

1. $5\frac{7}{8}$ 2. $5\frac{1}{12}$ 3. $6\frac{3}{4}$ 4. $1\frac{4}{5}$

5. $\frac{7}{18}$ 6. $3\frac{5}{12}$ 7. $3\frac{4}{15}$ 8. $\frac{17}{18}$

9. $59\frac{1}{4}$ 10. $91\frac{5}{6}$ 11. $8\frac{9}{10}$ 12. $17\frac{2}{3}$

WORKING WITH WORD PROBLEMS, PAGE 117

1. **(4)**
$$\begin{aligned}\frac{1}{6} &= \frac{2}{12}\\ \frac{1}{4} &= \frac{3}{12}\\ +\ \frac{1}{3} &= \frac{4}{12}\\ \hline &= \frac{9}{12} = \frac{3}{4}\end{aligned}$$

2. **(3)**
$$\begin{aligned}4\frac{1}{2} &= 4\frac{2}{4} = 3\frac{6}{4}\\ -\ 1\frac{3}{4} &= -1\frac{3}{4} = -1\frac{3}{4}\\ \hline & \qquad\qquad\qquad 2\frac{3}{4}\end{aligned}$$

3. **(2)**
$$\begin{aligned}3\frac{3}{4} &= 3\frac{15}{20}\\ +\ 2\frac{7}{10} &= 2\frac{14}{20}\\ \hline & \ \ 5\frac{29}{20} = 6\frac{9}{20}\end{aligned}$$

4. **(4)**
$$\begin{aligned}3\frac{1}{4} &= 3\frac{5}{20}\\ +\ 2\frac{1}{10} &= 2\frac{2}{20}\\ \hline & \ \ 5\frac{7}{20}\end{aligned}$$

5. **(2)**
$$\begin{aligned}2\frac{1}{2} &= 2\frac{4}{8}\\ 2\frac{1}{4} &= 2\frac{2}{8}\\ +\ 4\frac{3}{8} &= 4\frac{3}{8}\\ \hline & \ \ 8\frac{9}{8} = 9\frac{1}{8}\end{aligned}$$

6. **(1)**
$$\begin{aligned}189\frac{1}{4} &= 189\frac{2}{8} = 188\frac{10}{8}\\ -\ 15\frac{7}{8} &= 15\frac{7}{8} = 15\frac{7}{8}\\ \hline & \qquad\qquad\qquad 173\frac{3}{8}\end{aligned}$$

7. **(2)**
$$\begin{aligned}1\frac{1}{2} &= 1\frac{8}{16}\\ 1\frac{1}{8} &= 1\frac{2}{16}\\ \frac{9}{16} &= \frac{9}{16}\\ +\ 2\frac{1}{4} &= 2\frac{4}{16}\\ \hline & \ \ 4\frac{23}{16} = 5\frac{7}{16}\end{aligned}$$

$$\begin{aligned}12 &= 11\frac{16}{16}\\ -\ 5\frac{7}{16} &= 5\frac{7}{16}\\ \hline & \ \ 6\frac{9}{16}\end{aligned}$$

8. **(4)**

$$1 \tfrac{1}{2} = 1 \tfrac{4}{8}$$
$$3 \tfrac{1}{4} = 3 \tfrac{2}{8}$$
$$4 \tfrac{1}{8} = 4 \tfrac{1}{8}$$
$$+ \ 5 \tfrac{3}{8} = 5 \tfrac{3}{8}$$
$$\overline{\qquad 13 \tfrac{10}{8} = \ 14 \tfrac{2}{8} = \mathbf{14\tfrac{1}{4}}}$$

$$26 \tfrac{3}{4}$$
$$- \ 14 \tfrac{1}{4}$$
$$\overline{12 \tfrac{2}{4} = \mathbf{12\tfrac{1}{2}}}$$

9. **(1)**

$$2 \tfrac{1}{4} = 2 \tfrac{10}{40}$$
$$\tfrac{9}{10} = \quad \tfrac{36}{40}$$
$$1 \tfrac{7}{8} = 1 \tfrac{35}{40}$$
$$+ \quad \tfrac{3}{8} = \quad \tfrac{15}{40}$$
$$\overline{\qquad 3 \tfrac{96}{40} = 5 \tfrac{16}{40} = \mathbf{5\tfrac{2}{5}}}$$

$$9 \tfrac{3}{10} = 9 \tfrac{3}{10} = 8 \tfrac{13}{10}$$
$$- \ 5 \tfrac{2}{5} = 5 \tfrac{4}{10} = 5 \tfrac{4}{10}$$
$$\overline{\qquad\qquad\qquad\qquad \mathbf{3\tfrac{9}{10}}}$$

10. **(2)**

$$1 \tfrac{3}{8} = 1 \tfrac{9}{24}$$
$$\tfrac{3}{4} = \quad \tfrac{18}{24}$$
$$2 \tfrac{5}{12} = 2 \tfrac{10}{24}$$
$$+ \ 1 \tfrac{1}{8} = 1 \tfrac{3}{24}$$
$$\overline{\qquad 4 \tfrac{40}{24} = 5 \tfrac{16}{24} = \mathbf{5\tfrac{2}{3}}}$$

$$200 = 199 \tfrac{3}{3}$$
$$- \quad 5 \tfrac{2}{3} = \quad 5 \tfrac{2}{3}$$
$$\overline{\qquad\qquad\qquad \mathbf{194\tfrac{1}{3}}}$$

TRY IT, PAGE 119

1. $\tfrac{3}{8}$ 2. $\tfrac{1}{6}$ 3. $\tfrac{1}{4}$ 4. $\tfrac{2}{21}$ 5. $\tfrac{2}{7}$ 6. $\tfrac{1}{6}$

7. $\tfrac{9}{40}$ 8. $\tfrac{2}{9}$ 9. $\tfrac{3}{20}$ 10. $\tfrac{6}{35}$ 11. $\tfrac{1}{4}$ 12. $\tfrac{1}{8}$

13. $\tfrac{1}{6}$ 14. $\tfrac{3}{4}$ 15. $\tfrac{1}{10}$

TRY IT, PAGE 120

1. $\mathbf{1\tfrac{1}{4}}$ 2. $\tfrac{3}{10}$ 3. $\mathbf{3\tfrac{1}{3}}$ 4. $\mathbf{2\tfrac{2}{3}}$ 5. $\mathbf{6\tfrac{1}{8}}$ 6. $\mathbf{3\tfrac{1}{3}}$

7. $\mathbf{1\tfrac{1}{6}}$ 8. $\mathbf{2\tfrac{2}{5}}$ 9. $\mathbf{5\tfrac{1}{3}}$ 10. $\mathbf{3\tfrac{3}{5}}$ 11. $\tfrac{8}{9}$ 12. $\mathbf{4\tfrac{2}{3}}$

13. $\mathbf{1\tfrac{5}{7}}$ 14. $\mathbf{1\tfrac{1}{8}}$ 15. $\mathbf{2\tfrac{2}{5}}$

TRY IT, PAGE 121

1. $\frac{2}{9}$ 2. $\frac{3}{10}$ 3. $\frac{4}{25}$ 4. $\frac{4}{7}$ 5. $\frac{1}{9}$ 6. $\frac{1}{2}$

7. $\frac{8}{13}$ 8. $\frac{7}{30}$ 9. $\frac{1}{5}$ 10. $\frac{3}{14}$ 11. $\frac{3}{4}$ 12. $\frac{1}{4}$

13. $\frac{15}{22}$ 14. **1**   15. **1**

TRY IT, PAGE 122

1. **$4\frac{1}{8}$**   2. **$1\frac{19}{21}$**   3. **$4\frac{1}{8}$**   4. **$1\frac{13}{15}$**   5. **$1\frac{13}{27}$**   6. **3**

7. **$3\frac{1}{18}$**   8. **$1\frac{2}{5}$**   9. **$4\frac{1}{5}$**   10. **$1\frac{1}{12}$**   11. **$7\frac{1}{5}$**   12. **3**

13. $1\frac{1}{4}$ 14. **$9\frac{3}{8}$**   15. **$2\frac{1}{4}$**

WORKING WITH WORD PROBLEMS, PAGE 122

1. **(2)** $2\frac{1}{4} \times 7 = \frac{9}{4} \times \frac{7}{1} = \frac{63}{4} = \mathbf{15\frac{3}{4}}$

2. **(3)** $\frac{1}{\underset{1}{\cancel{2}}} \times \frac{\overset{1}{\cancel{2}}}{3} = \frac{1}{3}$

3. **(1)** $1\frac{1}{4} \times \frac{1}{2} = \frac{5}{4} \times \frac{1}{2} = \mathbf{\frac{5}{8}}$

4. **(2)** $6\frac{2}{3} \times \frac{2}{3} = \frac{20}{3} \times \frac{2}{3} = \frac{40}{9} = \mathbf{4\frac{4}{9}}$

5. **(2)** $350 \times \frac{2}{5} = \frac{\overset{70}{\cancel{350}}}{1} \times \frac{2}{\underset{1}{\cancel{5}}} = \frac{140}{1} = \mathbf{140}$

6. **(4)** $1\frac{1}{2} \times 1\frac{1}{3} = \frac{\overset{1}{\cancel{3}}}{\underset{1}{\cancel{2}}} \times \frac{\overset{2}{\cancel{4}}}{\underset{1}{\cancel{3}}} = \mathbf{2}$

7. **(1)** $5\frac{1}{2} \times 2\frac{1}{4} = \frac{11}{2} \times \frac{9}{4} = \frac{99}{8} = \mathbf{12}\frac{3}{8}$

8. **(2)** $1\frac{5}{8} \times 6 = \frac{13}{\underset{4}{8}} \times \frac{\overset{3}{6}}{1} = \frac{39}{4} = \mathbf{9}\frac{3}{4}$

9. **(4)** $1\frac{7}{10} \times \frac{5}{8} = \frac{17}{\underset{2}{10}} \times \frac{\overset{1}{5}}{8} = \frac{17}{16} = \mathbf{1}\frac{1}{16}$

10. **(3)** $\frac{\overset{1}{4}}{5} \times \frac{1}{2} \times \frac{3}{\underset{1}{4}} = \frac{3}{10}$

TRY IT, PAGE 124

1. **2**   2. $\mathbf{1}\frac{1}{6}$ 3. $\mathbf{1}\frac{1}{3}$ 4. $\mathbf{1}\frac{1}{8}$ 5. $\mathbf{1}\frac{3}{7}$ 6. $\frac{2}{3}$

7. $\mathbf{1}\frac{3}{5}$ 8. $\mathbf{2}\frac{3}{16}$ 9. $\frac{3}{5}$ 10. $\frac{15}{16}$ 11. $\mathbf{1}\frac{7}{9}$ 12. $\mathbf{1}\frac{2}{3}$

13. $\frac{35}{36}$ 14. $\frac{35}{36}$ 15. **1**

TRY IT, PAGE 125

1. $\frac{2}{9}$ 2. $\frac{4}{15}$ 3. **4**   4. $\frac{1}{12}$ 5. $\frac{7}{16}$ 6. $\frac{1}{4}$

7. $\mathbf{6}\frac{2}{3}$ 8. $\frac{2}{15}$ 9. **6**   10. $\mathbf{4}\frac{4}{5}$ 11. **30**   12. $\frac{1}{25}$

13. $\frac{3}{16}$ 14. $\frac{3}{28}$ 15. **25**

TRY IT, PAGE 126

1. **2**   2. **2**   3. **2**   4. $\frac{7}{10}$ 5. $\mathbf{1}\frac{1}{3}$ 6. $\frac{2}{3}$

7. $\frac{9}{14}$ 8. $\mathbf{2}\frac{1}{4}$ 9. **2**   10. $\frac{2}{3}$ 11. **6**   12. $\frac{1}{2}$

13. **1**   14. $\frac{1}{2}$ 15. $\mathbf{1}\frac{1}{2}$

TRY IT, PAGE 127, TOP

1. **7**　　2. **4$\frac{2}{3}$**　　3. **1**　　4. **4$\frac{4}{5}$**　　5. **9**　　6. **1$\frac{5}{9}$**

7. **3$\frac{1}{3}$**　　8. **1$\frac{5}{6}$**　　9. **3**　　10. **7$\frac{1}{2}$**　　12. **1$\frac{1}{4}$**　　12. **1$\frac{7}{9}$**

13. $\frac{7}{20}$　　14. **1**　　15. **5**

WORKING WITH WORD PROBLEMS, PAGE 127

1. **(4)**　$3 \div \frac{1}{8} = \frac{3}{1} \times \frac{8}{1} = $ **24**

2. **(2)**　$1\frac{1}{4} \div 5 = \frac{\cancel{5}^{1}}{4} \times \frac{1}{\cancel{5}_{1}} = \frac{1}{4}$

3. **(4)**　$18 \div 4\frac{1}{2} = \frac{\cancel{18}^{2}}{1} \times \frac{2}{\cancel{9}_{1}} = $ **4**

4. **(1)**　$3\frac{3}{8} \div 9 = \frac{\cancel{27}^{3}}{8} \times \frac{1}{\cancel{9}_{1}} = \frac{3}{8}$

$\frac{3}{8} \times 3 = \frac{3}{8} \times \frac{3}{1} = \frac{9}{8} = $ **1$\frac{1}{8}$**

5. **(2)**　$6\frac{1}{2} \div 1\frac{1}{2} = \frac{13}{\cancel{2}_{1}} \times \frac{\cancel{2}^{1}}{3} = \frac{13}{3} = 4\frac{1}{3}$ or **5 jars**

6. **(3)**　$\frac{5}{8} \div \frac{1}{4} = \frac{5}{\cancel{8}_{2}} \times \frac{\cancel{4}^{1}}{1} = \frac{5}{2} = $ **2$\frac{1}{2}$**

7. **(2)**　$2\frac{1}{3} \div \frac{7}{16} = \frac{\cancel{7}^{1}}{3} \times \frac{16}{\cancel{7}_{1}} = \frac{16}{3} = 5\frac{1}{3}$ or **5** whole burgers

8. **(4)**　$\frac{7}{8} \div 14 = \frac{\cancel{7}^{1}}{8} \times \frac{1}{\cancel{14}_{2}} = \frac{1}{16}$

9. **(1)**　$1\frac{7}{8} \div 5 = \frac{\cancel{15}^{3}}{8} \times \frac{1}{\cancel{5}_{1}} = \frac{3}{8}$

10. **(4)**　$5\frac{2}{3} \div 1\frac{8}{9} = \frac{\cancel{17}^{1}}{\cancel{3}_{1}} \times \frac{\cancel{9}^{3}}{\cancel{17}_{1}} = $ **3**

MEASURING YOUR PROGRESS, PAGE 129

TEST A

1. $\frac{3}{8}$ 2. $\frac{1}{5}$ 3. $\frac{1}{2}$ 4. **24$\frac{1}{4}$**

5. **1$\frac{7}{16}$**   6. $\frac{13}{24}$ 7. **15$\frac{1}{2}$**   8. **21**

9. **4$\frac{14}{15}$**   10. **1**   11. **68**   12. **57$\frac{5}{6}$**

13. **2$\frac{3}{4}$**   14. **18$\frac{31}{36}$**   15. **19$\frac{3}{20}$**   16. **2$\frac{3}{8}$**

17. **18$\frac{1}{5}$**   18. **5$\frac{5}{12}$**   19. **100$\frac{2}{9}$**   20. **5$\frac{6}{11}$**

21. **1**   22. **2**   23. $\frac{3}{8}$

24. **3$\frac{3}{8}$**   25. **2**   26. $\frac{1}{6}$

27. **4**   28. **1$\frac{3}{4}$**   29. **2**

30. **4$\frac{1}{2}$**   31. **2$\frac{1}{2}$**   32. **1**

33. $\frac{1}{5}$ 34. $\frac{8}{33}$ 35. **4**

36. $\frac{1}{12}$ 37. **3$\frac{3}{5}$**   38. **1$\frac{15}{34}$**

39. **4$\frac{4}{5}$**   40. **116$\frac{1}{4}$**

TEST B

1. **(1)** $1\frac{1}{4} + 2\frac{2}{4} = $ **3$\frac{3}{4}$**

2. **(3)** $4\frac{2}{3} - 3\frac{1}{3} = $ **1$\frac{1}{3}$**

3. **(1)** $\frac{\overset{10}{\cancel{50}}}{1} \times \frac{1}{\underset{1}{\cancel{5}}} = $ **10**

4. **(3)** $6 \div \frac{1}{3} = \frac{6}{1} \times \frac{3}{1} = \mathbf{18}$

5. **(2)** $2\frac{3}{5} + 3\frac{1}{5} = \mathbf{5}\frac{4}{5}$

6. **(2)** $135\frac{3}{4} - 124\frac{1}{4} = 11\frac{2}{4} = \mathbf{11}\frac{1}{2}$

7. **(2)** $4\frac{1}{8} + 2\frac{3}{8} + 5 = 11\frac{4}{8} = \mathbf{11}\frac{1}{2}$

8. **(1)** $72 \times \frac{1}{4} = \frac{\cancel{72}^{18}}{1} \times \frac{1}{\cancel{4}_1} = \mathbf{18}$ OR $72 \times \frac{3}{4} = \frac{\cancel{72}^{18}}{1} \times \frac{3}{\cancel{4}_1} = 54, \; 72 - 54 = \mathbf{18}$

9. **(1)** $2 \div \frac{1}{2} = \frac{2}{1} \times \frac{2}{1} = \mathbf{4}$

10. **(2)** $8 \div 2\frac{1}{3} = \frac{8}{1} \times \frac{3}{7} = \frac{24}{7} = 3\frac{3}{7}$ or **3 whole shelves**

11. **(2)**
$$2\frac{5}{6} = 2\frac{10}{12}$$
$$3\frac{1}{4} = 3\frac{3}{12}$$
$$+ \; 4\frac{2}{3} = 4\frac{8}{12}$$
$$9\frac{21}{12} = 10\frac{9}{12} = \mathbf{10}\frac{3}{4}$$

12. **(2)** $\frac{5}{8} \times \frac{1}{3} = \frac{\mathbf{5}}{\mathbf{24}}$

13. **(1)**
$$6\frac{1}{3} = 6\frac{2}{6} = 5\frac{8}{6}$$
$$- \; 1\frac{5}{6} = 1\frac{5}{6} = 1\frac{5}{6}$$
$$4\frac{3}{6} = \mathbf{4}\frac{1}{2}$$

14. **(3)** $\frac{3}{4} \div \frac{1}{16} = \frac{3}{\cancel{4}_1} \times \frac{\cancel{16}^4}{1} = \mathbf{12}$

15. **(3)** $11\frac{3}{10} \times 1\frac{1}{3} = \frac{113}{\cancel{10}_5} \times \frac{\cancel{4}^2}{3} = \frac{226}{15} = \mathbf{15}\frac{1}{15}$

16. **(2)**
$$4\tfrac{5}{6} = 4\tfrac{15}{18}$$
$$3\tfrac{2}{3} = 3\tfrac{12}{18}$$
$$+\ 2\tfrac{5}{9} = 2\tfrac{10}{18}$$
$$9\tfrac{37}{18} = \mathbf{11\tfrac{1}{18}}$$

17. **(4)** $3\tfrac{1}{3} \div \tfrac{5}{8} = \cancel{\tfrac{10}{3}}^{2} \times \dfrac{8}{\cancel{5}_{1}} = \tfrac{16}{3} = 5\tfrac{1}{3}$ or **5** whole burgers

18. **(2)** $\tfrac{5}{16} + \tfrac{5}{16} = \tfrac{10}{16} = \tfrac{5}{8}$

$$24\ = 23\tfrac{8}{8}$$
$$-\ \ \tfrac{5}{8} = \ \ \tfrac{5}{8}$$
$$\mathbf{23\tfrac{3}{8}}$$

19. **(1)** $\tfrac{2}{3} \div 8 = \dfrac{\cancel{2}^{1}}{3} \times \dfrac{1}{\cancel{8}_{4}} = \mathbf{\tfrac{1}{12}}$

20. **(2)** $3\tfrac{1}{7} \times 35 = \dfrac{22}{\cancel{7}_{1}} \times \dfrac{\cancel{35}^{5}}{1} = \mathbf{110}$

21. **(3)** $\tfrac{8}{1} \times \tfrac{1}{5} = \tfrac{8}{5} = 1\tfrac{3}{5}$

$$5\tfrac{1}{2} = 5\tfrac{5}{10} = 4\tfrac{15}{10}$$
$$-\ 1\tfrac{3}{5} = 1\tfrac{6}{10} = 1\tfrac{6}{10}$$
$$\mathbf{3\tfrac{9}{10}}$$

22. **(4)** $26\tfrac{2}{3} \div 6\tfrac{2}{3} = \dfrac{\cancel{80}^{4}}{\cancel{3}_{1}} \times \dfrac{\cancel{3}^{1}}{\cancel{20}_{1}} = \mathbf{4}$

23. **(3)** $2\tfrac{1}{2} \times \tfrac{3}{4} = \tfrac{5}{2} \times \tfrac{3}{4} = \tfrac{15}{8} = \mathbf{1\tfrac{7}{8}}$ OR $2\tfrac{1}{2} \times \tfrac{1}{4} = \tfrac{5}{2} \times \tfrac{1}{4} = \tfrac{5}{8}$

$$2\tfrac{1}{2} = 2\tfrac{4}{8} = 1\tfrac{12}{8}$$
$$-\ \ \tfrac{5}{8} = \ \ \tfrac{5}{8} = \ \ \tfrac{5}{8}$$
$$\mathbf{1\tfrac{7}{8}}$$

24. **(3)** $10\tfrac{1}{2} \div 14 = \dfrac{\cancel{21}^{3}}{2} \times \dfrac{1}{\cancel{14}_{2}} = \mathbf{\tfrac{3}{4}}$

Decimals

20. HOW MUCH DO YOU KNOW ALREADY?

Here are some problems with decimals. You will be asked to add, subtract, multiply, or change to fractions or mixed numbers. This will help you find out how much you know already. You will also find out what you need to study. Do as many problems as you can. Write your answers on the lines at the right. Then, turn to page 181. Check your answers. Page 181 will tell you what pages in the book to study.

1. $.78
    + .48

2. 312.65
        .07
    + 20.60

3. 142.25
    + 268.87

4. .33 + .1 + .015 =

5. 1.62 + .43 =

6. .3 + .9 =

7. .35
    − .12

8. 39.53
    − 6.71

9. $5.00
    − 2.98

10. .3 − .098 =

11. 2.03 − .10 =

12. 4 − 1.75 =

13. .1
    × .1

14. 4.1
    × 3.2

15. $21.06
    ×   5

1. _____
2. _____
3. _____
4. _____
5. _____
6. _____
7. _____
8. _____
9. _____
10. _____
11. _____
12. _____
13. _____
14. _____
15. _____

16. $5.63 \times 2.5 =$ 17. $3.20 \times 10 =$ 18. $.12 \times .8 =$

19. $5\overline{)3.5}$ 20. $5\overline{).415}$ 21. $8\overline{)7}$

22. $3.45\overline{)6.9}$ 23. $.75\overline{)1.5}$ 24. $.04\overline{)48}$

Change to decimals.

25. $\frac{3}{10}$ 26. $\frac{1}{2}$ 27. $\frac{1}{3}$

Change to fractions or mixed numbers.

28. $.625$ 29. 6.4 30. 1.05

16. _____
17. _____
18. _____
19. _____
20. _____
21. _____
22. _____
23. _____
24. _____
25. _____
26. _____
27. _____
28. _____
29. _____
30. _____

21. WHAT IS A DECIMAL?

Decimals are a special kind of fraction. Look at these examples:

.1	=	$\frac{1}{10}$	=	tenths
.01	=	$\frac{1}{100}$	=	hundredths
.001	=	$\frac{1}{1,000}$	=	thousandths
.0001	=	$\frac{1}{10,000}$	=	ten thousandths
.00001	=	$\frac{1}{100,000}$	=	hundred thousandths
.000001	=	$\frac{1}{1,000,000}$	=	millionths

The <u>decimal point</u> tells you that a number is a decimal fraction. The number of places in the decimal fraction tells you how much it is worth. Notice that the first place to the right of the decimal point is tenths. As you move to the right, the value <u>decreases</u> by multiples of 10.

Zeros help you put decimal fractions in the proper place. Look at these examples:

$$\frac{6}{100} = 6 \text{ hundredths} = .06$$

$$\frac{48}{1000} = 48 \text{ thousandths} = .048$$

$$\frac{7}{1000} = 7 \text{ thousandths} = .007$$

Try It (Answers on page 182.)

Write decimal fractions for these problems.

1. $\frac{8}{100}$ =   2.   3 thousandths =

3. $\frac{36}{1000}$ =   4.   4 hundredths =

5. $\frac{2}{1000}$ =   6.   15 thousandths =

149

7. $\frac{9}{1000}$ =   8.   6 thousandths =

9. $\frac{98}{1000}$ =   10.   2 hundredths =

11. $\frac{5}{1000}$ =   12.   38 thousandths =

A <u>mixed decimal</u> is a whole number and a decimal fraction. This is a mixed decimal:

whole number → 2.5 ← decimal fraction

Amounts of money are written as mixed decimals. Two dollars and thirteen cents is written as $2.13. It is two whole dollars and $\frac{13}{100}$ of a dollar.

All numbers to the *left* of the decimal point are whole numbers. All numbers to the *right* of the decimal point are decimal fractions.

Try It (Answers on page 182.)

Write these mixed numbers as mixed decimals.

1. $4\frac{9}{10}$ =   2. $104\frac{104}{1000}$ =

3. $23\frac{23}{1000}$ =   4. $8\frac{86}{100}$ =

5. $3\frac{3}{10}$ =   6. $1\frac{75}{100}$ =

7. $1\frac{6}{10}$ =   8. $15\frac{5}{10}$ =

9. $2\frac{3}{10}$ =   10. $99\frac{99}{100}$ =

COMPARING DECIMAL VALUES

You can compare the values of decimals by giving each decimal the same number of places. For example, to compare .14 and .3, first add a zero to the right of .3. Now compare .14 and .30. Both decimals have two places, or hundredths. Thirty hundredths is larger than fourteen hundredths. .3 is the larger of the two decimals. Notice that the zero added to .3 did not change the value of the decimal. The 3 is still in the tenths place. The zero simply makes the decimals easier to compare.

Try It (Answers on page 182.)

Circle the larger decimal in each pair.

1. .6 or .066

2. .04 or .044

3. .505 or .55

4. .2 or .12

5. .011 or .1

6. .3 or .321

7. .0024 or .002

8. .707 or .77

9. .68 or .068

10. .045 or .04

22. ADDING DECIMAL NUMBERS

Add decimal fractions as you add whole numbers. Line up the numbers so that the decimal points are under each other. Place a decimal point in the answer under all the other decimal points. Look at these examples:

$$
\begin{array}{r}
.33 \\
.21 \\
+\ .15 \\
\hline
.69
\end{array}
\qquad
.5 + .03 + .712 =
\begin{array}{r}
.500 \\
.030 \\
+\ .712 \\
\hline
1.242
\end{array}
$$

Notice that zeros are added as place holders in the second example. This does not change the value of the decimal fractions. The zeros help keep the columns straight.

Try It (Answers on page 182.)

1.
$$\begin{array}{r} .4 \\ +\ .1 \\ \hline \end{array}$$
2.
$$\begin{array}{r} .3 \\ +\ .6 \\ \hline \end{array}$$
3.
$$\begin{array}{r} .35 \\ +\ .43 \\ \hline \end{array}$$
4.
$$\begin{array}{r} .77 \\ +\ .11 \\ \hline \end{array}$$
5.
$$\begin{array}{r} .82 \\ +\ .17 \\ \hline \end{array}$$

6. $.1 + .008 =$ 7. $.128 + .06 =$ 8. $.33 + .019 =$

9. $.41 + .125 + .05 =$ 10. $.5 + .14 + .003 + .102 =$

ADDING WHOLE NUMBERS AND DECIMAL NUMBERS

Add mixed decimals just as you add decimal fractions. Line up the decimal points under each other. Whole numbers are to the left of the decimal point. Add a decimal point after a whole number that has no decimal fraction. Place a decimal point in your answer under all the other decimal points. Look at these examples:

$$
\begin{array}{r}
3.7 \\
+\ 1.2 \\
\hline
4.9
\end{array}
\qquad
\begin{array}{r}
1.62 \\
+\ .13 \\
\hline
1.75
\end{array}
\qquad
\begin{array}{r}
25.000 \\
.072 \\
+\ 1.121 \\
\hline
26.193
\end{array}
$$

Notice that zeros are added to the whole number 25 in the last example. This does not change the value of the number. The zeros hold the places to help you add the right columns.

Try It (Answers on page 182.)

1. $\begin{array}{r} 2.7 \\ + 6.2 \\ \hline \end{array}$

2. $\begin{array}{r} 10.3 \\ + 8.4 \\ \hline \end{array}$

3. $\begin{array}{r} 12.2 \\ + 12.2 \\ \hline \end{array}$

4. $\begin{array}{r} 7.8 \\ + 10.1 \\ \hline \end{array}$

5. $\begin{array}{r} 6.52 \\ + 3.47 \\ \hline \end{array}$

6. $\begin{array}{r} 1.76 \\ + 2.13 \\ \hline \end{array}$

7. $\begin{array}{r} 52.25 \\ + 2.21 \\ \hline \end{array}$

8. $\begin{array}{r} 103.42 \\ + 25.41 \\ \hline \end{array}$

9. $\begin{array}{r} 396.127 \\ + 2.352 \\ \hline \end{array}$

10. $\begin{array}{r} 1.835 \\ + .123 \\ \hline \end{array}$

11. $\begin{array}{r} 12.343 \\ 10.123 \\ + 14.323 \\ \hline \end{array}$

12. $\begin{array}{r} 76.224 \\ 11.111 \\ + 12.521 \\ \hline \end{array}$

13. $45 + 3.23 =$

14. $10 + .072 =$

15. $100 + 30.33 + .05 =$

CARRYING DECIMAL NUMBERS

Numbers are carried in decimal addition just as they are carried in whole number addition. Look at the examples below:

$\begin{array}{r} \overset{1}{4}.37 \\ + 2.25 \\ \hline 6.62 \end{array}$
$\begin{array}{r} \overset{1}{}.3 \\ + .9 \\ \hline 1.2 \end{array}$
$\begin{array}{r} \overset{1}{7}.95 \\ + 1.43 \\ \hline 9.38 \end{array}$

Try It (Answers on page 182.)

1. $\begin{array}{r} .4 \\ + .8 \\ \hline \end{array}$

2. $\begin{array}{r} .6 \\ + .7 \\ \hline \end{array}$

3. $\begin{array}{r} .9 \\ + .8 \\ \hline \end{array}$

4. $\begin{array}{r} .5 \\ + .6 \\ \hline \end{array}$

5. $\begin{array}{r} .7 \\ + .9 \\ \hline \end{array}$

6. 3.28
+ 1.35

7. 86.36
+ 5.47

8. 27.179
+ 3.019

9. 118.18
+ 18.18

10. 4.587
+ 3.274

11. 2.93
+ 3.42

12. 6.82
+ 5.44

13. 10.24
+ 6.94

14. 28.69
+ 14.75

15. 2.658
+ 1.848

ADDING MONEY

A zero and a decimal point make all the difference between a penny and a dollar. Look how each is written:

penny: .01
dollar: 1.00

A decimal point separates dollars from cents.

dollars → $23.63 ← cents

Decimal addition problems about money are added like other decimal problems. The decimal points must be lined up under each other. A dollar sign ($) is usually placed before the top number in the problem and in the answer. Look at the examples below:

$ .78
+   .48
$1.26

$1.95
+ 2.89
$4.84

$5 + 79¢ =

$5.00
+   .79
$5.79

Try It (Answers on page 183.)

1. $1.25
+ 3.45

2. $6.13
+ 4.19

3. $10.98
+ 6.95

4. $22.50
+ 13.25

5. $75.99
+ 23.49

6. $ .75
+   .25

7. $6.00
+ 6.00

8. $1.75
+   .02

9. $49.89
+ 1.75

10. $300.05
+ 10.02

11. $ .10
1.75
+   .55

12. $89.07
3.45
+ 27.81

13. $312.65 + 7¢ =   14. $10 + $1 + 10¢ =   15. $3.30 + 75¢ + $2 =

WORKING WITH WORD PROBLEMS

Here are some clue words that will help you solve addition problems. These clue words usually mean you must add.

1. **Added to**—The question might say, "Then $5 was added to the amount he had."

2. **Altogether**—The question might ask, "How many are there altogether?"

3. **How much** or **how many**—The question might ask, "How much money was spent?" Or, "How many miles did he drive?"

4. **Increased**—The question might say, "The bus fare was increased 15 cents."

5. **More**—The question might say, "He had to pay $5 more."

6. **Total** or **sum**—The question might ask, "What is the total amount of money spent?"

Read the Steps for Solving Word Problems on page xii again. Then work these problems. Find the correct answer.

1. Marge has an adjustable wrench. She opens it .6 of an inch to tighten a nut. It's too small, so she opens it .2 of an inch more. How many inches has she opened the wrench?
   (1) 4   (2) .4   (3) 8   (4) .8

2. Paul walks .3 of a mile to the Top Save grocery store. It is .1 of a mile farther to the drugstore. It is .2 of a mile more to the bakery. How many miles does Paul walk to the bakery?
   (1) .4   (2) .5   (3) .6   (4) .7

3. Mrs. Rivolta needs .9 of a yard of trim for the bottom of a dress. She also needs .2 of a yard for each sleeve. How many yards of trim does she need for the whole dress?
   (1) .9   (2) 1.3   (3) 1.1   (4) 1.9

4. Glenn spent $1.65 for a hamburger and $.70 for a root beer. There is $.14 sales tax. How much will he pay the cashier?
   (1) $1.86   (2) $2.39   (3) $3.75   (4) $2.49

5. An odometer counts the number of miles a car travels. On Saturday, Pat's odometer read 6,256.3 miles. He drives 68.8 miles on Sunday. How many miles does his odometer read now?
   (1) 63,251   (2) 6,325.1   (3) 632.51   (4) 6,251.1

6. Yesterday Elena spent $22.95 for a pair of sandals. She also spent $19.95 for a knit shirt and $14.95 for a blouse. There is a $4.05 sales tax on all of these items. How much did she spend altogether?
   (1) $59.90   (2) $61.70   (3) $61.90   (4) $62.35

7. The Olsons had their television set fixed. The bill had two sections—*Parts*: $18.29 for a new tube and $5.25 for a new dial; *Labor*: $45.00. What is the total amount of the bill?
   (1) $69.17   (2) $68.54   (3) $58.44   (4) $50.25

8. Last week John paid $9.50 for gas and $5.00 for oil. He also paid $4.75 to have his car washed and $6.25 to have a flat tire fixed. How much did John spend on his car last week?
   (1) $24.40   (2) $25.50   (3) $26.40   (4) $27.90

9. The first week of the month Gretchen spent $20.00 for food. The second week she spent $33.22. The third week she spent $29.89. The fourth week she spent $23.33. How much did she spend for food this month?
   (1) $95.34   (2) $95.55   (3) $104.66   (4) $106.44

10. A major league player had a batting average of .286. He increased it by .004. What is his new batting average?
    (1) .686   (2) .326   (3) .290   (4) .289

Check your answers on page 183.

23. SUBTRACTING DECIMAL NUMBERS

Subtract decimal numbers as you subtract whole numbers. Line up the numbers so that the decimal points are under each other. Place a decimal point in the answer under all the other decimal points. Look at these examples:

$$
\begin{array}{r} .9 \\ -\ .4 \\ \hline .5 \end{array}
\qquad
\begin{array}{r} .35 \\ -\ .12 \\ \hline .23 \end{array}
\qquad
12.486 - .23 =
\qquad
\begin{array}{r} 12.486 \\ -\ \ \ .230 \\ \hline 12.256 \end{array}
$$

Notice that a zero is used as a place holder in the last example.

Try It (Answers on page 183.)

1. $\begin{array}{r} .7 \\ -\ .3 \\ \hline \end{array}$
2. $\begin{array}{r} .8 \\ -\ .5 \\ \hline \end{array}$
3. $\begin{array}{r} .6 \\ -\ .1 \\ \hline \end{array}$
4. $\begin{array}{r} .44 \\ -\ .13 \\ \hline \end{array}$
5. $\begin{array}{r} .75 \\ -\ .22 \\ \hline \end{array}$

6. $\begin{array}{r} .742 \\ -\ .341 \\ \hline \end{array}$
7. $\begin{array}{r} .509 \\ -\ .306 \\ \hline \end{array}$
8. $\begin{array}{r} .895 \\ -\ .573 \\ \hline \end{array}$
9. $\begin{array}{r} 68.93 \\ -\ \ 4.91 \\ \hline \end{array}$
10. $\begin{array}{r} 74.989 \\ -\ \ 1.773 \\ \hline \end{array}$

11. $.89 - .51 =$
12. $.365 - .12 =$
13. $.672 - .6 =$

14. $14.6 - 3.2 =$
15. $29.87 - .75 =$
16. $5.895 - .063 =$

BORROWING IN SUBTRACTING DECIMALS

Sometimes you will need to borrow to subtract decimals. Borrow just as you do with whole numbers. Look at these examples:

$$
\begin{array}{r} \overset{8\ \ \ 15}{3\cancel{9}.\cancel{5}3} \\ -\ 16.71 \\ \hline 22.82 \end{array}
\qquad
\begin{array}{r} \overset{3\ 9\ 10}{1\cancel{4}.\cancel{0}\cancel{0}} \\ -\ \ \ .75 \\ \hline 13.25 \end{array}
\qquad
\begin{array}{r} \overset{2\ 9\ 9\ 15}{\cancel{3}\cancel{0}.\cancel{0}\cancel{5}} \\ -\ \ 9.67 \\ \hline 20.38 \end{array}
$$

Notice that zeros are added to the whole number 14 in the second example. This does not change the value of the number. The zeros hold the places so that you can subtract.

Try It (Answers on page 183.)

1.	6.2 − 3.3	2.	9.0 − 6.7	3.	12.8 − 1.9	4.	20.6 − 10.7

5.	312.6 − 1.7	6.	9.73 − 1.86	7.	7.54 − 2.98	8.	38.06 − 15.87

9.	43.00 − 27.49	10.	87.98 − 49.99	11.	40.004 − 22.407	12.	26.943 − 1.989

13. $152 - .076 =$ 14. $9.865 - 1.897 =$ 15. $5.04 - .56 =$

16. $10.05 - 1.98 =$ 17. $43 - 16.67 =$ 18. $140 - 1.008 =$

SUBTRACTING MONEY

Money problems are subtracted the same way as other decimal problems. The decimal points are lined up. A dollar sign ($) is usually placed before the top number and in the answer. Look at these examples:

$$
\begin{array}{r} \$.99 \\ - .64 \\ \hline \$.35 \end{array}
\qquad
\begin{array}{r} \$5.00 \\ - 2.98 \\ \hline \$2.02 \end{array}
$$

Try It (Answers on page 184.)

1.	$4.98 − 1.39	2.	$1.00 − .49	3.	$405.07 − 36.42	4.	$29.98 − 8.99	5.	$11.11 − 1.11

YOUR PAYCHECK

Your <u>gross pay</u> is the salary you earn every payday. It is the amount of money you earn before taxes, social security, and other <u>deductions</u> are taken out. Your <u>net pay</u>, or take-home pay, is the amount you take home after these deductions.

    To figure out your net pay, add up all your deductions. <u>Subtract</u> this amount from your gross pay. Look at this example:

Statement of Earnings and Deductions			
Gross Pay	$185.00	Net Pay	_____
Withholding Tax	$27.75	FICA	$13.87

(Income tax is called "Withholding Tax" and "FICA" is the deduction for social security benefits.)

    Step 1—Add up all the deductions: $27.75 + $13.87 = $41.62

    Step 2—Subtract the total deductions from the gross pay:
        $185.00 − $41.62 = $143.38, the net pay.

Try It (Answers on page 184.)

Find the net pay in each of these problems.

1. | Gross Pay | $142.35 | Net Pay | _____ |
   | Withholding Tax | $18.51 | FICA | $10.67 |

2. | Gross Pay | $179.94 | Net Pay | _____ |
   | Withholding Tax | $26.99 | FICA | $13.49 |

3. | Gross Pay | $224.02 | Net Pay | _____ |
   | Withholding Tax | $33.60 | FICA | $16.80 |

4. | Gross Pay | $270.00 | Net Pay | _____ |
   | Withholding Tax | $40.54 | FICA | $20.25 |

WORKING WITH WORD PROBLEMS

Here are some clue words that will help you solve subtraction problems. These clue words usually mean you must subtract.

1. **Decreased**—The question might ask, "By how much was the total decreased?"

2. **Difference**—The question might ask, "What is the difference in the prices?"

3. **Fewer** or **less**—The question might ask, "How many fewer seconds does it take Mary to run the quarter-mile?"

4. **How many more**—A problem may give you two numbers. It will ask how much larger one number is. Subtract the smaller number from the larger number. The question might ask, "How many more were sold on Saturday than were sold during the week?"

5. **Left** or **remains**—The question might ask, "How much is left?" Or, "How much remains?"

Read the Steps for Solving Word Problems on page xii again. Then work these problems. Find the correct answer.

1. Aurelio Rodriguez's batting average this year was .249. Last year it was .305. How much higher was his batting average last year?
   (1) .056   (2) .560   (3) .506   (4) .066

2. Super gasoline sells for $1.38 and 9/10 of one cent ($1.389) a gallon. Regular gasoline sells for $1.33 and 9/10 of one cent ($1.339) a gallon. How much cheaper is a gallon of regular gasoline than a gallon of super gasoline?
   (1) $.03   (2) $.04   (3) $.50   (4) $.05

3. Harry Jackson runs the 100-yard dash in 9.6 seconds. Jake Gadstand runs the same distance in 10 seconds. How many seconds slower is Jake Gadstand?
   (1) .4   (2) .04   (3) 1.4   (4) 1.6

4. A quart of milk costs $.64 at the local grocery store. The same quart of milk costs $.59 at the supermarket. How much is saved by buying the milk at the supermarket?
   (1) $.03   (2) $.04   (3) $.05   (4) $.005

5. Your gross pay is $241.31 per week. Your deductions are: federal income tax, $26.54; state income tax, $9.65; social security, $18.09; and health insurance, $.67. What is your net pay?
   (1) $186.36   (2) $187.03   (3) $187.46   (4) $214.77

6. Your monthly bills are: food, $219.36; rent, $325.00; transportation, $55.00; other expenses, $95.42. Your net pay is $871.29 per month. How much money do you have left after paying these bills?
   (1) $104.51   (2) $171.51   (3) $176.51   (4) $281.51

7. You plan to buy a blender that costs $29.95. If you buy it on credit, there is a service charge of $4.69. You decide to pay for it with a credit card. For the first credit payment, you put down $11.00. How much more do you still have to pay for the blender?
   (1) $14.26   (2) $15.69   (3) $18.95   (4) $23.64

8. The odometer on your car reads 47,060.0 miles. Last week it read 47,010.9 miles. How many miles have you traveled this week?
   (1) 50.1   (2) 59.1   (3) 49.1   (4) 60.1

9. You buy a record album for $8.98, tax included. If you give the cashier $10.00, how much change will you receive?
   (1) $1.02   (2) $2.00   (3) $2.02   (4) $3.02

10. At the supermarket, your check-out tape lists: meat, $4.39; soda, $1.99; bread, $1.25; and vegetables, $2.29. The cashier charges $.14 for sales tax. If you give the cashier $15.00, how much change will you get back?
   (1) $4.80   (2) $4.94   (3) $10.06   (4) $10.20

Check your answers on page 184.

24. MULTIPLYING DECIMAL NUMBERS

The decimal points do not have to be lined up to multiply decimal numbers.

In addition and subtraction of decimals, the decimal point is brought down to the answer. In multiplication of decimals, the number of decimal places in the answer must equal the number of decimal places in the problem. Look at the example below:

Addition:

```
   .40
   .40
 + .40
  1.20
```

Multiplication:

```
   .40   two decimal places
 ×   3   no decimal places
  1.20   two decimal places
```

Count the number of decimal places in the multiplication problem: 2 + 0 = 2. Since there are 2 decimal places in the problem, count off 2 decimal places from the right in the answer.

Count the number of decimal places in both numbers you are multiplying. Look at these examples:

```
   .5 one decimal place
 × 8 no decimal place
  4.0 one decimal place
```

```
  5.796 three decimal places
 ×   .8 one decimal place
 4.6368 four decimal places
```

Try It (Answers on page 184.)

1.
```
  .12
× 4
```

2.
```
  .34
× 5
```

3.
```
  .24
× 4
```

4.
```
  .75
× 8
```

5.
```
  .44
× 2
```

6.
```
  .67
× .2
```

7.
```
  .84
× .3
```

8.
```
  .26
× .6
```

9.
```
  .78
× .4
```

10.
```
  .92
× .2
```

11.
```
 3.25
×  .4
```

12.
```
 1.04
×  .2
```

13.
```
 50.3
×   3
```

14.
```
 2.125
×   5
```

15.
```
  .814
×   2
```

16.	15.94	17.	3.875	18.	5.66	19.	10.071	20.	.3675
	× 2		× 3		× .6		× .5		× .7

DECIMAL POINTS IN THE FINAL ANSWER

A decimal point is needed only in the final answer of a multiplication problem. The final answer has as many decimal places as there are in the problem. Look at this example:

$$
\begin{array}{r}
4.84 \\
\times\ 4.3 \\
\hline
1452 \\
1936\ \\
\hline
20.812
\end{array}
$$
two decimal places
one decimal place

three decimal places

Count off the number of decimal places: $2 + 1 = 3$. Since there are 3 decimal places in the problem, count off 3 decimal places from the right in the answer.

Try It (Answers on page 184.)

1.	63.2	2.	1.24	3.	424	4.	.47	5.	341.3
	× .64		× .12		× .26		× .85		× .08

6.	48.3	7.	5.67	8.	.279	9.	86.5	10.	149
	× 76		× 2.5		× .36		× .07		× .48

ZERO AS A PLACE HOLDER

There are 2 decimal places in the problem shown below. There must be 2 decimal places in the answer. Use a zero to hold a place. Two decimal places can then be counted off.

$$
\begin{array}{r}
.4 \\
\times\ .2 \\
\hline
.08
\end{array}
$$

Try It (Answers on page 185.)

1. .9 × .1	2. .3 × .2	3. .3 × .3	4. .4 × .4	5. .2 × .4
6. .16 × .4	7. .43 × .2	8. .23 × .3	9. .11 × .9	10. .22 × .2
11. .24 × .04	12. .06 × .03	13. .45 × .03	14. .39 × .07	15. .25 × .05
16. .35 × .23	17. .69 × .12	18. .25 × .31	19. .26 × .32	20. .106 × .24

MULTIPLYING MONEY

Amounts of money are multiplied the same way other decimal numbers are multiplied. The number of decimal places in the answer must equal the number of decimal places in the problem. Look at the example below;

$$\begin{array}{r} \$ \ 5.12 \\ \times \quad\quad 2 \\ \hline \$10.24 \end{array}$$

Try It (Answers on page 185.)

1. $46.98 × 2	2. $1.49 × 3	3. $9.99 × 7	4. $21.06 × 5	5. $69.99 × 8

6.	$.56	7.	$34.75	8.	$99.75	9.	$.97	10.	$27.82
	× 72		× 43		× 34		× 33		× 56

WORKING WITH WORD PROBLEMS

Here are some clue words that will help you solve multiplication problems. These clue words usually mean you must multiply.

1. **At**—The question might ask, "What is the cost of three dozen eggs at $1.02 a dozen?"

2. **How much** (or **how many**) **for a larger quantity**—The question might ask, "If candy costs $2.15 a pound, how much will seven pounds cost?"

3. **Times**—The question might say, "Five times as many gallons of paint were needed."

Read the Steps for Solving Word Problems on page xii again. Then work these problems. Find the correct answer.

1. Chicken is on sale for $.39 a pound. How much do four pounds of chicken cost?
   - (1) $.96
   - (2) $1.26
   - (3) $1.56
   - (4) $1.60

2. Record albums are on sale for $2.49 each. You buy 3 albums. How much change do you get from a $10.00 bill?
   - (1) $2.49
   - (2) $2.53
   - (3) $7.51
   - (4) $7.47

3. The horse "Lucked Out" paid $3.20 for every dollar bet on him. If you made a $5.00 bet, how much money would you win?
   - (1) $15.00
   - (2) $16.00
   - (3) $150.00
   - (4) $160.00

4. You earn $3.35 an hour at a part-time job. Monday you work 4 hours. Tuesday you work 5 hours. Wednesday you don't work. Thursday you work 6 hours. Friday you work 4 hours. How much money did you earn for the week?

   (1) $63.65   (2) $44.25   (3) $59.00   (4) $56.05

5. Some people get paid by the hour. The money they earn is sometimes called "straight time." Overtime pay is often called "time and a half." This means that for one hour's overtime work, you will get paid your regular hourly pay plus an additional half of that. ("Time and a half" can be written with decimals as 1.5.) You make $3.92 an hour straight time. If you are paid time and a half for overtime, how much is your hourly overtime pay?

   (1) $.588   (2) $5.88   (3) $58.80   (4) $588.00

6. A mother receives $32.92 each week from public assistance for each of her children. She has 7 children. How much money does she receive each week?

   (1) $160.44   (2) $230.44   (3) $232.34   (4) $254.34

7. One bus ticket costs $.55. A 20-trip book of commuter tickets costs $9.50. If you took 20 trips, how much did you save by buying the book of commuter tickets?

   (1) $.55   (2) $1.50   (3) $1.55   (4) $2.50

8. The bus fare is raised from $.55 to $.80. You take the bus twice a day, Monday through Friday. How much more per week do you now spend on bus fare?

   (1) $1.25   (2) $1.50   (3) $2.50   (4) $2.25

9. You get 16.5 miles on one gallon of gas. How many miles can you travel on 12.5 gallons of gas?

   (1) 165.5   (2) 206.25   (3) 2062.5   (4) 577.5

10. A plant employs 563 workers, each of whom earns $177.42 a week. If the workers get a cost of living wage increase of $9.24 a week, what is the total cost to management of the workers' weekly wages?

    (1) $105,089.58   (2) $99,887.46   (3) $94,685.34   (4) $5,202.12

Check your answers on page 185.

25. DIVIDING DECIMAL NUMBERS

What is the difference between these two division problems?

$$3\overline{)9} \qquad 3\overline{)\overset{.3}{.9}}$$

In the first problem, a whole number is being divided. In the second problem, a decimal number is being divided. Notice where the decimal point is placed in the answer. The decimal point in the answer always goes <u>directly</u> <u>above</u> the decimal point in the problem.

Try It (Answers on page 186.)

1. $2\overline{).8}$ 2. $3\overline{).6}$ 3. $6\overline{).66}$ 4. $4\overline{).84}$ 5. $3\overline{).93}$

6. $3\overline{)3.9}$ 7. $2\overline{)8.4}$ 8. $4\overline{)4.4}$ 9. $6\overline{)12.6}$ 10. $5\overline{)15.5}$

11. $8\overline{)7.28}$ 12. $6\overline{)4.86}$ 13. $9\overline{)5.49}$ 14. $7\overline{)6.37}$ 15. $4\overline{)2.48}$

ZEROS AS PLACE HOLDERS

Zeros are used to hold decimal places when you cannot divide. Look at the example below:

$$
\begin{array}{r}
.083 \\
5\overline{).415} \\
\underline{40} \\
15 \\
\underline{15}
\end{array}
$$

You cannot divide 5 into 4. Put a zero above the 4. It acts as a place holder. Then divide 5 into 41.

Try It (Answers on page 186.)

1. $6\overline{).366}$ 2. $3\overline{).219}$ 3. $8\overline{).568}$ 4. $7\overline{).287}$ 5. $2\overline{).164}$

6. 4)‾.028‾   7. 2)‾.012‾   8. 6)‾.024‾   9. 9)‾.063‾   10. 7)‾.021‾

ADDING ZEROS TO DIVIDE

Adding a decimal point and a zero after a whole number will not change the value of that number. You can add any number of zeros you wish *after* the decimal point. The value of the whole number will be the same.

Sometimes adding zeros helps in division. Look at the example below:

	Step 1	Step 2	Check
5)‾3‾	5)‾3.0‾	$\begin{array}{r}.6\\5\overline{)3.0}\end{array}$	$\begin{array}{r}.6\\\times\ 5\\\hline 3.0\end{array}$

Step 1—To divide this problem, add a decimal point and a zero.

Step 2—Divide: $30 \div 5 = 6$. The decimal point in the answer goes directly above the decimal point in the problem. Notice that the 6 is placed *after* the decimal point. It is a decimal number.

Try It (Answers on page 186.)

1. 6)‾3‾   2. 5)‾4‾   3. 15)‾3‾   4. 10)‾6‾   5. 12)‾6‾

6. 5)‾6‾   7. 6)‾9‾   8. 5)‾8‾   9. 15)‾21‾   10. 25)‾45‾

GETTING RID OF REMAINDERS

Sometimes you can get rid of remainders in decimal problems. Add zeros and keep dividing until the problem works out evenly. The added zeros do not change the value of the decimal fraction. The fraction $\frac{3}{10}$ is the same as $\frac{30}{100}$ or $\frac{300}{1,000}$. Look at this example:

Problem: Solution: Check:

$$
\begin{array}{r}
.975 \\
8\overline{)7.800} \\
7\ 2 \\
\hline
60 \\
56 \\
\hline
40 \\
40 \\
\hline
00
\end{array}
\qquad
\begin{array}{r}
.975 \\
\times\quad 8 \\
\hline
7.800
\end{array}
$$

$8\overline{)7.8}$

Try It (Answers on page 186.)

1. $8\overline{)2.8}$ 2. $5\overline{).38}$ 3. $4\overline{)26}$ 4. $6\overline{).51}$ 5. $7\overline{)1.407}$

6. $4\overline{)4.2}$ 7. $5\overline{)3.1}$ 8. $24\overline{)3}$ 9. $8\overline{)3}$ 10. $16\overline{)1}$

ROUNDING OFF DECIMAL PLACES

Sometimes you will find that you have a remainder no matter how many zeros you add and divide into. In this case, <u>round off</u> your answer. You can round off an answer to any place you want. The examples below show decimals rounded off to the thousandths (third) place.

To round off a decimal number to the third place, look at the fourth-place number. If the fourth-place number is 4 or less, leave the third-place number as it is. Drop all numbers to the right of the third-place number. Look at the examples below:

.4932 rounded off becomes .493
.6744 rounded off becomes .674

If the fourth-place number is 5 or more, add 1 to the third-place number. Drop all numbers to the right of the third-place number. Look at the examples below:

.3645 rounded off becomes .365
.1286 rounded off becomes .129

Use the same process to round off to any place.

Try It (Answers on page 186.)

Divide. Round off your answers to three places

1. 7)8.1   2. 6)49.3   3. 7).58

4. 6)2   5. 3)11   6. 8)2.51

DIVIDING BY DECIMAL NUMBERS

You can divide by a decimal number. First, you must change the decimal number to a whole number. Look at the example below:

	Step 1	Step 2	Step 3	Check
.8).48	.8.).48	8.).4.8	.6 8)4.8	.8 ×.6 .48

Step 1—Change .8 to the whole number 8. To do this, move the decimal point one place to the right: .8 —► 8.

Step 2—Because the decimal point is moved to the right outside the division box, move the decimal point the same number of places to the right inside the division box: .48 —► 4.8

Step 3—Divide: 4.8 ÷ 8 = .6

Try It (Answers on page 186.)

1. $.2\overline{).8}$
2. $.6\overline{)4.2}$
3. $.9\overline{).81}$
4. $.7\overline{)5.6}$

5. $.12\overline{).48}$
6. $1.8\overline{).72}$
7. $1.6\overline{)9.6}$
8. $.43\overline{).344}$

9. $.18\overline{).3708}$
10. $2.3\overline{)1.656}$
11. $6.7\overline{)40.2}$
12. $3.9\overline{)138.84}$

MORE ON ADDING ZEROS TO DIVIDE

Sometimes you must add zeros to the number inside the division box before you divide. Look at the example below:

	Step 1	Step 2	Step 3	Check
$3.45\overline{)6.9}$	$3.45.\overline{)6.9}$	$345.\overline{)6.90.}$	$345\overline{)690.}^{\quad 2.}$	$\begin{array}{r} 3.45 \\ \times\quad 2 \\ \hline 6.90 \end{array}$

Step 1—Change 3.45 to the whole number 345 by moving the decimal point two places to the right.

Step 2—Move the decimal point the same number of places, or 2 places, inside the division box. Add a zero so that you can move the decimal point 2 places: 6.9 ⟶ 690.

Step 3—Divide: $690 \div 345 = 2$.

Sometimes you may need to add more than 1 zero. Look at the examples below:

$.134.\overline{)6.700.}$ 2 zeros added

$.125.\overline{)79.000.}$ 3 zeros added

Notice that a decimal point must be placed after the whole number 79 in the second example before it can be moved.

Try It (Answers on page 186.)

1. $3.45\overline{)6.9}$ 2. $1.34\overline{)6.7}$ 3. $.125\overline{)79.}$ 4. $.75\overline{)1}$

5. $.013\overline{)3.9}$ 6. $.22\overline{)1.1}$ 7. $7.5\overline{)60}$ 8. $1.86\overline{)9.3}$

9. $.04\overline{)5}$ 10. $4.2\overline{)861}$ 11. $5.64\overline{)84.6}$ 12. $.722\overline{)25.27}$

DIVIDING MONEY

Amounts of money are divided the same way other decimal numbers are divided. The decimal point in the answer always goes above the decimal point in the problem. Look at the example below:

$$\frac{\$\ .30}{4\overline{)\$1.20}}$$

Try It (Answers on page 186.)

1. $25\overline{)\$1.25}$ 2. $16\overline{)\$17.60}$ 3. $32\overline{)\$80.00}$ 4. $12\overline{)\$202.20}$

5. $10\overline{)\$1,200.00}$ 6. $17\overline{)\$212.50}$ 7. $5\overline{)\$65.55}$ 8. $4\overline{)\$128.96}$

9. $365\overline{)\$438.00}$ 10. $35\overline{)\$100.45}$ 11. $312\overline{)\$780.00}$ 12. $12\overline{)\$9,126.60}$

CHANGING FRACTIONS TO DECIMALS

Fractions can be made into decimals by dividing the numerator (top) by the denominator (bottom). Look at the example below.

Change $\frac{3}{5}$ to a decimal.

Step 1	Step 2	Step 3
$\frac{3}{5} = 5\overline{)3}$	$\begin{array}{r} .6 \\ 5\overline{)3.0} \\ \underline{3\ 0} \\ 0 \end{array}$	$\frac{3}{5} = .6$

Step 1—Set up the division problem. Divide the numerator by the denominator.

Step 2—Add a decimal point and a zero to divide. $30 \div 5 = 6$.

Step 3— $\frac{3}{5}$ expressed as a decimal is **.6**.

Sometimes you must add several zeros before the division comes out even. For example, change $\frac{5}{8}$ to a decimal.

Step 1	Step 2	Step 3
$\frac{5}{8} = 8\overline{)5}$	$\begin{array}{r} .625 \\ 8\overline{)5.000} \\ \underline{4\ 8} \\ 20 \\ \underline{16} \\ 40 \\ \underline{40} \end{array}$	$\frac{5}{8} = .625$

Step 1—Divide the numerator by the denominator.

Step 2—Add a decimal point and a zero. Divide. You will need to add two more zeros to come out even.

Step 3— $\frac{5}{8}$ expressed as a decimal is **.625**.

Some divisions never come out even, no matter how many zeros are added. For example, change $\frac{1}{3}$ to a decimal.

Step 1	Step 2	Step 3
$\frac{1}{3} = 3\overline{)1}$	$\begin{array}{r} .333 \\ 3\overline{)1.000} \\ \underline{9} \\ 10 \\ \underline{9} \\ 10 \\ \underline{9} \\ 1 \end{array}$	$\frac{1}{3} = .333$

Step 1—Divide the numerator by the denominator.

Step 2—Add a zero and divide. The remainder is 1. Add another zero and divide again. Try it a third time. You will always have a remainder of one.

Step 3—You can stop after the third place and write $\frac{1}{3}$ as **.333**.

No matter how far you carry this division, you still get <u>repeating</u> threes. You can stop the division and write three dots (. . .) to show the number continues on:

$$.333 \ldots$$

Or, you can put the remainder in fraction form:

$$.333 = .33\tfrac{1}{3}$$

Divisions like $\frac{1}{3}$ are called <u>repeating</u> <u>decimals</u>.

Try It (Answers on page 187.)

Change these fractions to decimals. Use the fractional form for remainders in repeating decimals.

1. $\frac{1}{5}$ 2. $\frac{1}{2}$ 3. $\frac{1}{4}$ 4. $\frac{3}{4}$ 5. $\frac{4}{5}$

6. $\frac{2}{3}$ 7. $\frac{3}{8}$ 8. $\frac{1}{6}$ 9. $\frac{7}{8}$ 10. $\frac{2}{9}$

CHANGING DECIMALS TO FRACTIONS

To change decimals to fractions:

Step 1—Drop the decimal point. Write the decimal numerals as the top of the fraction.

Step 2—The bottom of the fraction will be a 1 followed by the number of zeros equal to the number of places in the original decimal.

Step 3—Reduce the fraction if possible.

Look at the examples below:

$$.35 \quad = \quad \frac{35}{100} \quad = \quad \frac{7}{20}$$

$$.007 \quad = \quad \frac{7}{1,000}$$

$$.008 \quad = \quad \frac{8}{1,000} \quad = \quad \frac{1}{125}$$

Try It (Answers on page 187.)

Change these decimals to fractions. Reduce to lowest terms if possible.

1. .3   2. .66   3. .75   4. .004   5. .0001

WORKING WITH WORD PROBLEMS

Here are some clue words that will help you solve division problems. These clue words usually mean you must divide.

1. **Divide**—The question might say, "The money was divided equally among four people."

2. **Each**—The question might ask, "How much does each one cost?"

Read the Steps for Solving Word Problems on page xii again. Then work these problems. Find the correct answer.

1. Rita drove 204.8 miles in 4 hours. How many miles did she travel each hour?
   (1) 51.2   (2) 52.1   (3) 521   (4) 51.1

2. Rick is grocery shopping. He can buy 5 cans of tomato soup for $1.00. How much is one can?
   (1) $.19   (2) $.20   (3) $.21   (4) $.22

3. Norma rides the subway to work. Subway tokens are $.90 each. How many tokens can she get for $15.30?
   (1) 15   (2) 16   (3) 17   (4) 18

4. Marian called her aunt long distance. There was a charge of $.78 for additional units. Additional units are charged at $.03 per minute. How many extra minutes did she talk?
   (1) 24   (2) 25   (3) 26   (4) 27

5. Gasoline is $.84 and 9/10 of a cent a gallon ($.849). Phyllis spent $16.98 to fill her gas tank. How many gallons of gas did she put in?
   (1) 17   (2) 18   (3) 19   (4) 20

6. Roger bought a package of cheese marked $\frac{3}{8}$ pound at the grocery store. He bought a package marked .625 pound at the Cheese Shop. How many pounds of cheese did he buy altogether?
   (1) 3.292   (2) 1   (3) .1   (4) .000001

7. A shipment of boxes of chocolates weighs 360 pounds. Each box weighs 1.25 pounds. How many boxes are being shipped?
   (1) 450   (2) 347   (3) 288   (4) 2.88

8. The door of a safe is .912 foot thick. The door is make up of 12 equal layers of steel. What is the thickness of each layer?
   (1) .076   (2) .0076   (3) 7.6   (4) 76

9. Fred paid $3.45 for 5.75 feet of lumber. What was the cost per foot?
   (1) $1.67   (2) $1.65   (3) $.77   (4) $.60

10. There are 28.4 grams in one ounce. How many ounces are there in 171.11 grams?
    (1) .6025   (2) 6.025   (3) 60.25   (4) 602.5

Check your answers on page 187.

26. MEASURING YOUR PROGRESS

TEST A

Add or Subtract

1.	.4 .8 + .9	2.	4.57 2.93 + 4.87	3.	2.467 3.56 + .987	4.	34.98 2.73 25.68 + .98

5.	396.1 426.3 891.5 + 271.5	6.	50.025 16.125 1.625 + 2.5	7.	$12.98 3.29 1.05 + .75	8.	$100.00 8.75 13.20 + 8.05

9. $25 + .025 + 1.2 =$

10. $\$14 + \$1.98 + 5¢ =$

11.	.8 − .4	12.	5.4 − 2.9	13.	.679 − .398	14.	8.000 − 5.975

15.	176.29 − 94.49	16.	101.7 − 23.025	17.	$10.05 − 1.98	18.	$15.00 − 6.75

19. $16.231 − .05 =$

20. $.9654 − .128 =$

Multiply or Divide

21.
```
   1.29
×     6
```

22.
```
   .37
× .3
```

23.
```
   2.53
× .06
```

24.
```
   12.63
×    .5
```

25.
```
   .39
× .04
```

26.
```
   4.693
×   .18
```

27.
```
   468.39
×    .27
```

28.
```
   $15.95
×      5
```

29.
```
   $105.98
×       17
```

30.
```
   .034
× .78
```

31. 8)4.88

32. 12)11.316

33. 7).0497

34. .4)35.6

35. .012).00060

36. .56)2.4472

37. .15)45

38. 5)$15.95

39. 149)$372.50

40. .125)50

TEST B

Read the Steps for Solving Word Problems on page xii again. Then work these problems. Find the correct answer.

1. The annual rainfall last year was: New York, 42.94 inches; Chicago, 33.0 inches; and Houston, 45.74 inches. What was the total rainfall for the three cities?
   (1) 142.94   (2) 125.74   (3) 121.68   (4) 131.78

2. Mr. Wilson earns $7,200.00 annually (yearly). He is paid on a monthy basis. How much does he earn each month?
   (1) $600.00   (2) $500.00   (3) $700.00   (4) $650.00

3. Mark has saved $34.39 for a coat that costs $49.00. How much more money does he need?
(1) $15.51   (2) $14.61   (3) $13.72   (4) $12.81

4. Last month Mr. Henrik, owner of a drugstore, made bank deposits (put money in the bank) of $269.43, $179.63, $476.59, and $191.74. He made withdrawals (took money out of the bank) of $121.89, $263.00, $97.99, $139.58, and $6.66. Before the deposits and withdrawals, his bank balance was $497.35. Find his new bank balance.
(1) $489.54   (2) $985.62   (3) $1,614.74   (4) $1,125.20

5. Mr. Jacob has driven 792.8 miles in his new car. How much farther does he have to go before his 1,000 mile check-up?
(1) 108.3   (2) 108.8   (3) 207.2   (4) 209.4

6. Louise bought some new clothes on sale. Her dress cost $35.98; shoes, $25.50; hat, $15.25; and bag, $18.75. How much did she spend?
(1) $93.38   (2) $95.03   (3) $95.39   (4) $95.48

7. How much will Dennis pay for 8.6 gallons of paint thinner at $.33 a gallon?
(1) $2.83   (2) $2.84   (3) $2.93   (4) $3.84

8. At $.17 per pound, how many pounds of apples can you buy for $1.02?
(1) 5   (2) 7   (3) 6   (4) 8

9. Jim and Glenna are buying new linoleum for their kitchen. Find the cost of 15 square yards of linoleum at $3.30 per square yard.
(1) $56.43   (2) $59.70   (3) $49.50   (4) $39.50

10. The school's baseball team bought 12 balls at $2.25 each and 9 bats at $2.45 each. What was the total amount spent?
(1) $50.95   (2) $49.05   (3) $39.00   (4) $49.50

11. Marty bought three packages of cheese at the store. One was marked 1.875 pounds, one was marked $\frac{3}{4}$ pound, and one was marked $1\frac{1}{8}$ pound. How many pounds of cheese did he buy altogether?
(1) 10.5   (2) 4.333   (3) 3.075   (4) 3.75

12. Six people contributed $127.50 each to the Midtown Community Center. The Center wants to divide the money equally among 17 of its clients. How much will each client get?
(1) $7.50   (2) $11.59   (3) $21.25   (4) $45

13. Subway tokens are 95¢ each. Wanda wants to buy 12 tokens to last for a week. She gives the cashier $20 for the tokens. How much change will she get?
(1) $8.60   (2) $9.60   (3) $11.40   (4) $17.25

14. Rita paid $94.50 for 5.25 yards of brocade. What was the cost per yard?
(1) $18   (2) $1.80   (3) $89.25   (4) $496.13

15. Walter went to the grocery store. He bought a box of aluminum foil for $1.09, a package of chicken breasts for $1.96, a can of tomato sauce for 54¢, and a carton of cottage cheese for $1.19. The cashier charged him 38¢ sales tax. How much change will he get if he pays with a $10 bill?
(1) $6.14   (2) $5.94   (3) $5.84   (4) $4.84

Check your answers on page 188.

ANSWERS

HOW MUCH DO YOU KNOW ALREADY?, PAGE 147

Put an X by each problem you missed. Put an X by each problem you did not do. The problems are listed under the titles of the chapters in this book. If you have <u>one or more Xs</u> in a section, study that chapter to improve your skills.

ADDING DECIMAL NUMBERS (pages 152-156)

1. **$1.26**   2. **333.32**   3. **411.12**   4. **.445**   5. **2.05**   6. **1.2**

SUBTRACTING DECIMAL NUMBERS (pages 157-161)

7. **.23**   8. **32.82**   9. **$2.02**   10. **.202**   11. **1.93**   12. **2.25**

MULTIPLYING DECIMAL NUMBERS (pages 162-166)

13. **.01**   14. **13.12**   15. **$105.30**   16. **14.075**   17. **32**   18. **.096**

DIVIDING DECIMAL NUMBERS (pages 167-176)

19. **.7**   20. **.083**   21. **.875**   22. **2**   23. **2**   24. **1200**

25. **.3**   26. **.5**   27. **.33$\frac{1}{3}$**   28. **$\frac{5}{8}$**   29. **6$\frac{2}{5}$**   30. **1$\frac{1}{20}$**

TRY IT, PAGE 149

1. **.08**   2. **.003**   3. **.036**   4. **.04**
5. **.002**   6. **.015**   7. **.009**   8. **.006**
9. **.098**   10. **.02**   11. **.005**   12. **.038**

TRY IT, PAGE 150

1. **4.9**   2. **104.104**   3. **23.023**   4. **8.86**
5. **3.3**   6. **1.75**   7. **1.6**   8. **15.5**
9. **2.3**   10. **99.99**

TRY IT, PAGE 151

1. **.6**   2. **.044**
3. **.55**   4. **.2**
5. **.1**   6. **.321**
7. **.0024**   8. **.77**
9. **.68**   10. **.045**

TRY IT, PAGE 152

1. **.5**   2. **.9**   3. **.78**   4. **.88**   5. **.99**
6. **.108**   7. **.188**   8. **.349**   9. **.585**   10. **.745**

TRY IT, PAGE 153, TOP

1. **8.9**   2. **18.7**   3. **24.4**   4. **17.9**   5. **9.99**
6. **3.89**   7. **54.46**   8. **128.83**   9. **398.479**   10. **1.958**
11. **36.789**   12. **99.856** 13. **48.23**   14. **10.072**   15. **130.38**

TRY IT, PAGE 153, BOTTOM

1. **1.2**   2. **1.3**   3. **1.7**   4. **1.1**   5. **1.6**
6. **4.63**   7. **91.83**   8. **30.198**   9. **136.36**   10. **7.861**
11. **6.35**   12. **12.26**   13. **17.18**   14. **43.44**   15. **4.506**

TRY IT, PAGE 154

1. **$4.70**	2. **$10.32**	3. **$17.93**	4. **$35.75**	5. **$99.48**
6. **$1.00**	7. **$12.00**	8. **$1.77**	9. **$51.64**	10. **$310.07**
11. **$2.40**	12. **$120.33**	13. **$312.72**	14. **$11.10**	15. **$6.05**

WORKING WITH WORD PROBLEMS, PAGE 155

1. **(4)** .6
   + .2
    .8

2. **(3)** .3
   .1
  + .2
   .6

3. **(2)** .9
   .2
  + .2
  1.3

4. **(4)** $1.65
   .70
  + .14
  $2.49

5. **(2)** 6,256.3
  + 68.8
  6,325.1

6. **(3)** $22.95
   19.95
   14.95
  + 4.05
  $61.90

7. **(2)** $18.29
   5.25
  + 45.00
  $68.54

8. **(2)** $ 9.50
   5.00
   4.75
  + 6.25
  $25.50

9. **(4)** $ 20.00
   33.22
   29.89
  + 23.33
  $106.44

10. **(3)** .286
  + .004
  .290

TRY IT, PAGE 157

1. **.4**	2. **.3**	3. **.5**	4. **.31**	5. **.53**
6. **.401**	7. **.203**	8. **.322**	9. **64.02**	10. **73.216**
11. **.38**	12. **.245**	13. **.072**		
14. **11.4**	15. **29.12**	16. **5.832**		

TRY IT, PAGE 158, TOP

1. **2.9**	2. **2.3**	3. **10.9**	4. **9.9**
5. **310.9**	6. **7.87**	7. **4.56**	8. **22.19**
9. **15.51**	10. **37.99**	11. **17.597**	12. **24.954**
13. **151.924**	14. **7.968**	15. **4.48**	
16. **8.07**	17. **26.33**	18. **138.992**	

TRY IT, PAGE 158, BOTTOM

1. **$3.59**   2. **$.51**   3. **$368.65**   4. **$20.99**   5. **$10.00**

TRY IT, PAGE 159

1. **$113.17**   2. **$139.46**   3. **$173.62**   4. **$209.21**

WORKING WITH WORD PROBLEMS, PAGE 160

1. **(1)**
```
   .305
 − .249
   .056
```

2. **(4)**
```
  $1.389
 − 1.339
  $  .05
```

3. **(1)**
```
   10.0
 −  9.6
    .4
```

4. **(3)**
```
   $.64
 −  .59
   $.05
```

5. **(1)**
```
  $26.54        $241.31
    9.65      −  54.95
   18.09        $186.36
 +   .67
  $54.95
```

6. **(3)**
```
  $219.36        $871.29
   325.00      − 694.78
    55.00        $176.51
 +  95.42
  $694.78
```

7. **(4)**
```
  $29.95        $34.64
 +  4.69      − 11.00
  $34.64        $23.64
```

8. **(3)**
```
   47,060.0
 − 47,010.9
      49.1
```

9. **(1)**
```
   $5.00
 − 3.98
   $1.02
```

10. **(2)**
```
  $ 4.39        $15.00
    1.99      − 10.06
    1.25        $ 4.94
    2.29
 +   .14
  $10.06
```

TRY IT, PAGE 162

1. **.48**	2. **1.7**	3. **.96**	4. **6**	5. **.88**
6. **.134**	7. **.252**	8. **.156**	9. **.312**	10. **.184**
11. **1.3**	12. **.208**	13. **150.9**	14. **10.625**	15. **1.628**
16. **31.88**	17. **11.625**	18. **3.396**	19. **5.0355**	20. **.25725**

TRY IT, PAGE 163

1. **40.448**	2. **.1488**	3. **110.24**	4. **.3995**	5. **27.304**
6. **3670.8**	7. **14.175**	8. **.10044**	9. **6.055**	10. **71.52**

TRY IT, PAGE 164, TOP

1. **.09**	2. **.06**	3. **.09**	4. **.16**	5. **.08**
6. **.064**	7. **.086**	8. **.069**	9. **.099**	10. **.044**
11. **.0096**	12. **.0018**	13. **.0135**	14. **.0273**	15. **.0125**
16. **.0805**	17. **.0828**	18. **.0775**	19. **.0832**	20. **.02544**

TRY IT, PAGE 164, BOTTOM

1. **$93.96**	2. **$4.47**	3. **$69.93**	4. **$105.30**	5. **$559.92**
6. **$40.32**	7. **$1,494.25**	8. **$3,391.50**	9. **$32.01**	10. **$1,557.92**

WORKING WITH WORD PROBLEMS, PAGE 165

1. **(3)**
$$
\begin{array}{r}
\$\ .39 \\
\times\ \ \ .\ 4 \\
\hline
\mathbf{\$1.56}
\end{array}
$$

2. **(2)**
$$
\begin{array}{r}
\$2.49 \\
\times\ \ \ \ \ 3 \\
\hline
\$7.47
\end{array}
\qquad
\begin{array}{r}
\$10.00 \\
-\ \ \ 7.47 \\
\hline
\mathbf{\$\ \ 2.53}
\end{array}
$$

3. **(2)**
$$
\begin{array}{r}
\$3.20 \\
\times\ \ 5.00 \\
\hline
\mathbf{\$16.00}
\end{array}
$$

4. **(4)**
$$
\begin{array}{r}
4 \\
5 \\
6 \\
+\ 4 \\
\hline
19
\end{array}
\qquad
\begin{array}{r}
\$\ 3.35 \\
\times\ \ \ \ 19 \\
\hline
30\ 15 \\
33\ 50 \\
\hline
\mathbf{\$63.65}
\end{array}
$$

5. **(2)**
$$
\begin{array}{r}
\$\ \ 3.92 \\
\times\ \ \ \ \ 1.5 \\
\hline
1\ 9\ 60 \\
3\ 9\ 20 \\
\hline
\mathbf{\$5.8\ 80}
\end{array}
$$

6. **(2)**
$$
\begin{array}{r}
\$\ 32.92 \\
\times\ \ \ \ \ \ 7 \\
\hline
\mathbf{\$230.44}
\end{array}
$$

7. **(2)**
$$
\begin{array}{r}
\$\ \ .55 \\
\times\ \ \ \ 20 \\
\hline
\$11.00
\end{array}
\qquad
\begin{array}{r}
\$11.00 \\
-\ \ 9.50 \\
\hline
\mathbf{\$\ 1.50}
\end{array}
$$

8. **(3)**
$$
\begin{array}{r}
\$.80 \\
-\ .55 \\
\hline
\$.25
\end{array}
\qquad
\begin{array}{r}
\$\ .25 \\
\times\ \ \ 10 \\
\hline
\mathbf{\$2.50}
\end{array}
$$

9. **(2)**
$$
\begin{array}{r}
1\ 6.5 \\
\times\ \ 1\ 2.5 \\
\hline
8\ 2\ 5 \\
33\ 0\ 0 \\
165\ 0\ 0 \\
\hline
\mathbf{206.2\ 5}
\end{array}
$$

10. **(1)**
$$
\begin{array}{r}
\$177.42 \\
+\ \ \ 9.24 \\
\hline
\$186.66
\end{array}
\qquad
\begin{array}{r}
\$\ \ \ \ 186.66 \\
\times\ \ \ \ \ \ 5\ 63 \\
\hline
559\ 98 \\
11\ 199\ 60 \\
93\ 330\ 00 \\
\hline
\mathbf{\$105,089.58}
\end{array}
$$

TRY IT, PAGE 167, TOP

1. **.4**	2. **.2**	3. **.11**	4. **.21**	5. **.31**
6. **1.3**	7. **4.2**	8. **1.1**	9. **2.1**	10. **3.1**
11. **.91**	12. **.81**	13. **.61**	14. **.91**	15. **.62**

TRY IT, PAGE 167, BOTTOM

1. **.061**	2. **.073**	3. **.071**	4. **.041**	5. **.082**
6. **.007**	7. **.006**	8. **.004**	9. **.007**	10. **.003**

TRY IT, PAGE 168

1. **.5**	2. **.8**	3. **.2**	4. **.6**	5. **.5**
6. **1.2**	7. **1.5**	8. **1.6**	9. **1.4**	10. **1.8**

TRY IT, PAGE 169

1. **.35**	2. **.076**	3. **6.5**	4. **.085**	5. **.201**
6. **1.05**	7. **.62**	8. **.125**	9. **.375**	10. **.0625**

TRY IT, PAGE 170

1. **1.157**	2. **8.217**	3. **.083**
4. **.333**	5. **3.667**	6. **.314**

TRY IT, PAGE 171

1. **4**	2. **7**	3. **.9**	4. **8**
5. **4**	6. **.4**	7. **6**	8. **.8**
9. **2.06**	10. **.72**	11. **6**	12. **35.6**

TRY IT, PAGE 172, TOP

1. **2**	2. **5**	3. **632**	4. **1.33$\frac{1}{3}$**
5. **300**	6. **5**	7. **8**	8. **5**
9. **125**	10. **205**	11. **15**	12. **35**

TRY IT, PAGE 172, BOTTOM

1. **$.05**	2. **$1.10**	3. **$2.50**	4. **$16.85**
5. **$120.00**	6. **$12.50**	7. **$13.11**	8. **$32.24**
9. **$1.20**	10. **$2.87**	11. **$2.50**	12. **$760.55**

TRY IT, PAGE 174

1. **.2**   2. **.5**   3. **.25**   4. **.75**   5. **.8**

6. **.66 $\frac{2}{3}$**   7. **.375**   8. **.16 $\frac{2}{3}$**   9. **.875**   10. **.22 $\frac{2}{9}$**

TRY IT, PAGE 175

1. $\frac{3}{10}$ 2. $\frac{33}{50}$ 3. $\frac{3}{4}$ 4. $\frac{1}{250}$ 5. $\frac{1}{10,000}$

WORKING WITH WORD PROBLEMS, PAGE 176

1. **(1)** $4\overline{)204.8}$ = **51.2**

2. **(2)** $5\overline{)\$1.00}$ = **$.20**

3. **(3)** $\$.90\overline{)\$15.30}$ = **17**

4. **(3)** $.03\overline{)\$.78}$ = **26**

5. **(4)** $.849\overline{)16.980}$ = **20.**

6. **(2)** $\frac{3}{8} = 8\overline{)3.000}$ = **.375**

$$\begin{array}{r} .625 \\ + \ .375 \\ \hline 1.000 \end{array}$$

7. **(3)** $1.25\overline{)360.00}$ = **288.**

8. **(1)** $12\overline{).912}$ = **.076**

9. **(4)** $5.75\overline{)\$3.45\,00}$ = **.60**

10. **(2)** $28.4\overline{)171.1\,100}$ = **6.025**

MEASURING YOUR PROGRESS, PAGE 177

Test A

1. **2.1**	2. **12.37**	3. **7.014**	4. **64.37**
5. **1,985.4**	6. **70.275**	7. **$18.07**	8. **$130.00**
9. **26.225**	10. **$16.03**	11. **.4**	12. **2.5**
13. **.281**	14. **2.025**	15. **81.80**	16. **78.675**
17. **$8.07**	18. **$8.25**	19. **16.181**	20. **.8374**
21. **7.74**	22. **.111**	23. **.1518**	24. **6.315**
25. **.0156**	26. **.84474**	27. **126.4653**	28. **$79.75**
29. **$1,801.66**	30. **.02652**	31. **.61**	32. **.943**
33. **.0071**	34. **89**	35. **.05**	36. **4.37**
37. **300**	38. **$3.19**	39. **$2.50**	40. **400**

Test B

1. **(3)**
```
    42.94
    33.0
 +  45.74
   121.68
```

2. **(1)**
```
        $600.00
12)$7,200.00
```

3. **(2)**
```
   $49.00
 − 34.39
   $14.61
```

4. **(2)**
```
$   497.35
    269.43
    179.63
    476.59
 +  191.74
 $1,614.74
(previous balance + deposits)
```

```
 $121.89
  263.00
   97.99
  139.58
 +  6.66
 $629.12
(withdrawals)
```

```
 $1,614.74
 −  629.12
   $985.62
```

5. **(3)**
```
   1,000.0
 −   792.8
     207.2
```

6. **(4)**
```
   $35.98
    25.50
    15.25
 +  18.75
   $95.48
```

7. **(2)**
```
$    .3 3
×     8.6
    1 9 8
  2 6 4 0
 $2.8 3 8
```
2.838 rounded off to the nearest cent = **$2.84**

8. **(3)**
```
            6.
.17.)$1.02.
```

9. **(3)**
```
$  3.30
×    .15
  16 50
  33 0
 $49.50
```

10. **(2)**
```
$  2.25
×    .12
   4 50
  22 5
 $27.00
```

```
$  2.45
×    .9
 $22.05
```

```
  $27.00
 + 22.05
  $49.05
```

11. **(4)**
```
          .75
3/4 = 4)3.00
```

```
          .125
1/8 = 8)1.000
```

```
   1.875
    .750
 + 1.125
   3.75
```

12. **(4)** $127.50
     ×   6
        $765.00

                   $45.00
               17)$765.00

13. **(1)** $ .95
     ×   12
         1 80
         9 50
        $11.40

        $20.00
     − 11.40
     $ 8.60

14. **(1)**   **18.**
      5.25)94.50

15. **(4)** $1.09
        1.96
         .54
        1.19
     +   .38
        $5.16

        $10.00
     −   5.16
     $4.84

Percents

27. HOW MUCH DO YOU KNOW ALREADY?

Here are some problems with percents. You will be asked to change percents to decimals or fractions. You will also be asked to find interest. This will help you find out how much you know already. You will also find out what you need to study. Do as many problems as you can. Write your answers on the lines at the right. Then, turn to page 227. Check your answers. Page 227 will tell you what pages in the book to study.

1. Shade 25%   2. Shade 50%   3. Shade 75%

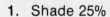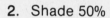

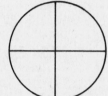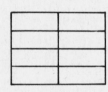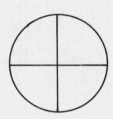

Change to decimals.

4. 6%   5. 123%   6. 12.5%

Change to percents.

7. .03   8. .76   9. .007

Change to fractions. Reduce where possible.

10. 7%   11. 9%   12. 20%

13. 75%   14. ¾%   15. 4%

1. _____
2. _____
3. _____
4. _____
5. _____
6. _____
7. _____
8. _____
9. _____
10. _____
11. _____
12. _____
13. _____
14. _____
15. _____

Find the percents.

16. 2% of 100 =   17. 8% of 150 =   18. 5% of 620 =

19. 25% of 600 =   20. $33\frac{1}{3}$% of 312 =   21. $12\frac{1}{2}$% of 728 =

Fill in the blanks.

22. 10 = ____% of 40   23. 35 = ____% of 175

24. 75 = ____% of 30   25. 8 = 25% of ____

26. 6 = 200% of ____   27. 12 = 75% of ____

28. Find the interest on $600 at 4% for 2 years.

29. Find the total repaid for a loan of $800 at 6% for 6 months.

16. ——————
17. ——————
18. ——————
19. ——————
20. ——————
21. ——————
22. ——————
23. ——————
24. ——————
25. ——————
26. ——————
27. ——————
28. ——————
29. ——————

28. WHAT IS A PERCENT?

Percent means hundredth. That is, percent is another way of writing a fraction with a denominator of 100. The symbol for percent is %. Look at the examples below:

$$\frac{7}{100} = 7\%$$

$$\frac{23}{100} = 23\%$$

$$\frac{3}{10} = \frac{30}{100} = 30\%$$

In the last example, both numerator and denominator were multiplied by 10. This changes the fraction to hundredths. The number can then be written as a percent.

Two place decimals also represent hundredths. Look at these examples:

$$.07 = 7\%$$

$$.23 = 23\%$$

$$.3 = .30 = 30\%$$

SHOWING PERCENTS

Fractions, decimals, and percents are all ways of showing a part of a whole. The whole is 100% or $\frac{100}{100}$. Look at the example below.

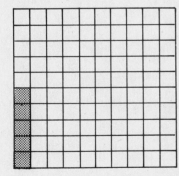

The large square is divided into 100 small squares. Each small square is 1% (or $\frac{1}{100}$ or .01) of the large square. 5 small squares are shaded. This means that 5% (or $\frac{5}{100}$ or .05) of the large square is shaded. 95 small squares are not shaded. This means that 95% (or $\frac{95}{100}$ or .95) of the large square is not shaded.

Try It (Answers on page 228.)

Look at the figures. Fill in the blanks.

1.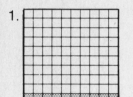

% shaded _____
% not shaded _____
total _____

2.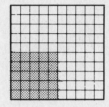

% shaded _____
% not shaded _____
total _____

3.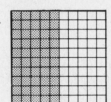

% shaded _____
% not shaded _____
total _____

4.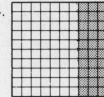

% shaded _____
% not shaded _____
total _____

5.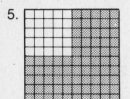

% shaded _____
% not shaded _____
total _____

6.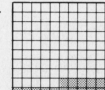

% shaded _____
% not shaded _____
total _____

7.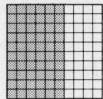

% shaded _____
% not shaded _____
total _____

8.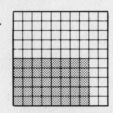

% shaded _____
% not shaded _____
total _____

CHANGING PERCENTS TO DECIMALS

Since percent means hundredths, a decimal can be used instead. Look at these examples:

$$1\% = .01$$
$$23\% = .23$$
$$85\% = .85$$

Here is the rule for changing percents to decimals.

Step 1	Step 2	Step 3
5%	5.	.05

Step 1—Remove the % sign.

Step 2—Move the decimal point two places to the left. If the point is not written, it is at the right end.

Step 3—Put zeros in front if necessary.

No matter how many places in a percent or decimal, the rule is the same. Look at the examples below:

$$200\% = 2.00 \text{ or } 2$$

$$125\% = 1.25$$

Try It (Answers on page 228.)

Change these percents to decimals.

1. 15% =	2. 37% =	3. 63% =	4. 9% =	5. 67% =
6. 10% =	7. 12.5% =	8. 6% =	9. 3% =	10. 99% =
11. 100% =	12. 350% =	13. 175% =	14. 250% =	15. 300% =

CHANGING DECIMALS TO PERCENTS

Change a decimal to a percent by reversing the rule for changing percents to decimals.

Step 1	Step 2	Step 3
.03	03%	3%

Step 1—Move the decimal point two places to the right. If the decimal point is at the end, drop it.

Step 2—Add % sign.

Step 3—Drop any zeros in front of the percent.

Look at these examples:

$$.25 = 25\% \qquad 1.03 = 103\% \qquad .375 = 37.5\%$$

Try It (Answers on page 228.)

Change these decimals to percents.

1. .03 =   2. .2 =   3. .75 =   4. .6 =   5. .19 =

6. .007 =   7. .075 =   8. .0825 =   9. .90 =   10. .67 =

FRACTIONAL PERCENTS

To change a fractional percent to a decimal, first change the fraction to a decimal (see page 29). Then, follow the rules for changing a percent to a decimal. Look at these examples:

$$\frac{1}{2}\% = .5\% = .005$$

$$\frac{1}{4}\% = .25\% = .0025$$

Try It (Answers on page 228.)

Change these fractional percents to decimals.

1. $\frac{3}{4}\% =$ 2. $\frac{1}{10}\% =$ 3. $\frac{1}{5}\% =$ 4. $\frac{3}{10}\% =$ 5. $\frac{9}{10}\% =$

6. $\frac{1}{2}\% =$ 7. $\frac{1}{4}\% =$ 8. $\frac{4}{5}\% =$ 9. $\frac{3}{5}\% =$ 10. $\frac{2}{5}\% =$

CHANGING PERCENTS TO FRACTIONS

Percents can also be rewritten as fractions. Remember percent means hundredths. Look at the example below:

Step 1	Step 2	Step 3	Step 4
50%	$\underline{50}$	$\frac{50}{100}$	$\frac{50 \div 50}{100 \div 50} = \frac{1}{2}$

To change a percent to a fraction:

Step 1—Remove the percent sign.

Step 2—Write the numeral for the numerator.

Step 3—Write 100 for the denominator.

Step 4—Reduce to lowest terms if possible.

Try It (Answers on page 229.)

Change these percents to fractions. Reduce your answers to lowest terms.

1. 9% =   2. 8% =   3. 20% =   4. 35% =   5. 86% =

6. 75% =   7. 40% =   8. 5% =   9. 80% =   10. 6% =

CHANGING FRACTIONS TO PERCENTS

To change a fraction into a percent:

Step 1—Change the fraction into a decimal. That is, divide the numerator by the denominator.

Step 2—Follow the rule for changing a decimal into a percent.

Look at these examples:

$$\frac{3}{4} = 4\overline{)3.00} \quad \begin{array}{r} .75 \\ \hline \end{array} \quad = 75\%$$

$$\begin{array}{r} 2\ 8 \\ \hline 20 \\ 20 \\ \hline \end{array}$$

$$\frac{5}{8} = 8\overline{)5.000} \quad \begin{array}{r} .625 \\ \hline \end{array} \quad = 62.5\%$$

$$\begin{array}{r} 4\ 8 \\ \hline 20 \\ 16 \\ \hline 40 \\ 40 \\ \hline \end{array}$$

Try It (Answers on page 229.)

Change these fractions to percents.

1. $\frac{1}{5} =$ 2. $\frac{1}{2} =$ 3. $\frac{1}{8} =$ 4. $\frac{1}{20} =$ 5. $\frac{1}{4} =$

6. $\frac{3}{8} =$ 7. $\frac{1}{3} =$ 8. $\frac{2}{5} =$ 9. $\frac{7}{8} =$ 10. $\frac{2}{3} =$

COMMON PERCENTS AND EQUIVALENTS

Here is a table of some common percents and their equivalents. Study the table. It will save you time and make some percent problems easier.

Percent	Fraction	Decimal	Percent	Fraction	Decimal
1%	$\frac{1}{100}$	.01	40%	$\frac{2}{5}$	.4
5%	$\frac{1}{20}$	.05	50%	$\frac{1}{2}$	.5
10%	$\frac{1}{10}$	.1	60%	$\frac{3}{5}$	.6
$12\frac{1}{2}\%$	$\frac{1}{8}$	.125	$62\frac{1}{2}\%$	$\frac{5}{8}$	.625
$16\frac{2}{3}\%$	$\frac{1}{6}$	.1666...	$66\frac{2}{3}\%$	$\frac{2}{3}$	.666...
20%	$\frac{1}{5}$	.2	70%	$\frac{7}{10}$	.7
25%	$\frac{1}{4}$	.25	75%	$\frac{3}{4}$	.75
30%	$\frac{3}{10}$	.3	80%	$\frac{4}{5}$	.8
$33\frac{1}{3}\%$	$\frac{1}{3}$	.333...	$87\frac{1}{2}\%$	$\frac{7}{8}$	.875
$37\frac{1}{2}\%$	$\frac{3}{8}$	.375	90%	$\frac{9}{10}$	.9
			100%	$\frac{1}{1}$	1

Try It (Answers on page 229.)

Complete this table.

	Percent	Fraction	Decimal
1.	150%		
2.			.25
3.		$\frac{1}{2}$	
4.	75%		
5.		$\frac{1}{10}$	
6.			.2
7.	30%		
8.			.35
9.		$\frac{13}{20}$	
10.	62.5%		

WORKING WITH WORD PROBLEMS

Read the Steps for Solving Word Problems on page xii again. Then work these problems. Find the correct answer.

1. A dollar is worth 100 pennies. What percent of a dollar is 50 pennies?

   (1) $\frac{1}{2}$%　　(2) .50%　　(3) 50%　　(4) .05%

2. A dollar is worth 10 dimes. What percent of a dollar is 3 dimes?

   (1) 3%　　(2) $\frac{1}{3}$%　　(3) 300%　　(4) 30%

3. A meter contains 100 centimeters. A shelf which is 65 centimeters long is what percent of a meter?

   (1) 65%　　(2) 6.5%　　(3) .65%　　(4) $6\frac{1}{2}$%

4. Fred's batting average for the baseball season was .347. Express Fred's batting average as a percent.

   (1) 347%　　(2) 3.47%　　(3) 34.7%　　(4) .347%

5. There are 10 students in Mr. Adams's night school math class. On Thursday two of the students were absent. What percent of the students were absent?

   (1) 2%　　(2) 20%　　(3) 200%　　(4) .2%

6. Mavis has finished 75% of the book she is reading. What fraction of the book has she finished?
   (1) $\frac{7}{5}$ (2) $\frac{5}{7}$ (3) $\frac{3}{4}$ (4) $\frac{4}{3}$

7. Mark is buying a stereo on an installment plan. He paid 20% on his balance. What fraction of the balance does he still owe?
   (1) $\frac{4}{5}$ (2) $\frac{1}{5}$ (3) $\frac{4}{50}$ (4) $\frac{1}{50}$

8. In a recent election $\frac{9}{20}$ of the registered voters went to the polls to vote. What percent of the registered voters did not vote?
   (1) 55%   (2) 45%   (3) 92%   (4) 29%

9. Roger finished $\frac{1}{16}$ of his test. What percent of the test did he complete?
   (1) 16%   (2) 6.25%   (3) 62.5%   (4) .0625%

10. Carlos worked Monday through Friday. He completed $\frac{1}{8}$ of a job each work day. What percent of the job was left to finish?
    (1) 67.5%   (2) 65.7%   (3) 37.5%   (4) 3.75%

Check your answers on page 229.

29. FINDING A PERCENT OF A NUMBER

To find a percent of a number, change the percent to a decimal or fraction and multiply. Look at these examples for finding 25% of 40.

Step 1	Step 2	Step 3
$25\% = \frac{1}{4}$	$\frac{1}{\cancel{4}} \times \frac{\cancel{40}^{10}}{1} = 10$	25% of $40 = 10$

Step 1—Change the percent to a fraction.

Step 2—Multiply the number by the fraction.

Step 3—10 is 25% of 40.

Step 1	Step 2	Step 3
$25\% = .25$	$.25 \times 40 = 10.00$	25% of $40 = 10$

Step 1—Change the percent to a decimal.

Step 2—Multiply the number by the decimal.

Step 3—10 is 25% of 40.

In some cases it is easier to use the decimal form. For example, to find 24% of 150:

$$24\% = .24$$

$$
\begin{array}{r}
1\,50 \\
\times \quad .24 \\
\hline
6\,00 \\
30\,0 \\
\hline
36.00 \\
\end{array}
$$

$$24\% \text{ of } 150 = 36$$

Try It (Answers on page 230.)

Solve. Use fractions or decimals.

1. 60% of 80 =   2. 50% of 128 =   3. 16% of 150 =

4. $66\frac{2}{3}$% of 96 =   5. 25% of 240 =   6. 8% of 350 =

7. 90% of 500 =   8. 150% of 8 =   9. 200% of 8 =

10. 250% of 8 =   11. .5% of 400 =   12. .25% of 600 =

ESTIMATING ANSWERS

A good way to check whether an answer is reasonable is to <u>estimate</u>. Fractions can be very helpful in estimating answers. Look at this example.

Find 48% of 400.

48% is approximately 50%, or $\frac{1}{2}$.

The answer is approximately $\frac{1}{2}$ of 400, or 200.

We know the actual answer is a little less than 200, since 48% is a little less than 50%. We have a good idea of what the answer is.

Try It (Answers on page 230.)

Solve. First, *estimate* the answer. Then solve for the *exact* answer.

	Fraction Used for Estimate	Estimated Answer	Actual Answer
1. Find 74% of 200	_____	_____	_____
2. Find 9% of 300	_____	_____	_____
3. Find 24% of 96	_____	_____	_____
4. Find 49% of 86	_____	_____	_____
5. Find 32% of 9	_____	_____	_____

PERCENT APPLICATIONS

Many practical applications of percent involve two-step operations.

Step 1—Find the amount.

Step 2—Add or subtract the amount. Watch for key words that tell you to add or subtract.

Look at the following examples. You will find these types of problems in the "Working With Word Problems" exercise.

Increase

Firemen in South Bay received an 8% pay raise this year. Last year they made $20,500. How much will South Bay firemen make this year?

Step 1—Find the amount of in-   Step 2—Increases are added.
          crease.

$20,500   original salary $20,500   original salary
×   .08   percentage raise   + 1,640   raise
$1,640   amount of raise $22,140   new salary

The firemen will make $22,140 this year.

Markup

Fashion Find Shoppes must cover store expenses and make a profit. They add a 50% markup to all items they sell. They will buy their new summer dresses at $18 each. What will they sell the dresses for in their shops?

Step 1—Find the amount of markup.

$$
\begin{array}{rl}
\$\ 18 & \text{cost} \\
\times\quad .50 & \text{percentage markup} \\
\hline
\$9.00 & \text{amount of markup}
\end{array}
$$

Step 2—Markups are added.

$$
\begin{array}{rl}
\$18 & \text{cost} \\
+\quad 9 & \text{amount of markup} \\
\hline
\$27 & \text{selling price}
\end{array}
$$

The selling price of the dresses will be $27.

Decrease

The number of smokers in Addison Adult Center has decreased by 5% since last year. Last year there were 420 smokers. How many Addison students smoke this year?

Step 1—Find the amount of decrease.

$$
\begin{array}{rl}
420 & \text{original number} \\
\times\quad .05 & \text{percentage decrease} \\
\hline
21.00 & \text{decrease}
\end{array}
$$

Step 2—Decreases are subtracted.

$$
\begin{array}{rl}
420 & \text{original number} \\
-\quad 21 & \text{decrease} \\
\hline
399 & \text{number of smokers now}
\end{array}
$$

There are 399 smokers at Addison this year.

Discount

Up Towne Stores is having a sale. $25 sweaters are discounted 15%. What is the sale price of the sweaters?

Step 1—Find the amount of discount.

$$
\begin{array}{rl}
\$\ 25 & \text{original price} \\
\times\quad .15 & \text{discount rate} \\
\hline
\$3.75 & \text{amount of discount}
\end{array}
$$

Step 2—Discounts are subtracted.

$$
\begin{array}{rl}
\$25.00 & \text{original price} \\
-\quad 3.75 & \text{discount} \\
\hline
\$21.25 & \text{sale price}
\end{array}
$$

The sweaters are on sale for $21.25.

Depreciation

Machinery in Crompton Mills depreciates (loses value) 20% a year due to use and wear. Drilling machines cost $4,000 new. How much are they worth after one year?

Step 1—Find the amount of depreciation.

Step 2—Depreciation is subtracted.

$4,000	new cost		$4,000	cost new
× .20	depreciation rate		− 800	depreciation
$ 800	amount of depreciation in 1 year		$3,200	depreciated value

The machines are worth $3,200 after one year.

WORKING WITH WORD PROBLEMS

Read the Steps for Solving Word Problems on page xii again. Then work these problems. Find the correct answer.

1. Salesclerks at Clark's Carpet City receive a 10% commission on sales. The new sales clerk sold $829 worth of carpets this week. What was his commission?
   (1) $8.29   (2) $82.90   (3) $.82   (4) $10

2. Zimbels sells boys' suits for $84. They are on sale this Easter for 25% off. How much can you save this Easter?
   (1) $25   (2) $8.40   (3) $63   (4) $21

3. Rich has moved to a state with a 6% sales tax on all purchases. He bought a ladder for $56. How much sales tax will he have to pay?
   (1) $6   (2) $33.60   (3) $3.36   (4) $59.36

4. A new car depreciates about 20% the first year. Anna has just bought a new car for $8,000. How much will it depreciate the first year?
   (1) $1,600   (2) $200   (3) $160   (4) $400

5. The population of Springfield has increased $33\frac{1}{3}$% since last year. Last year the population was 9,000. How much has the population increased?
(1) 297   (2) 3,000   (3) 12,000   (4) 2,970

6. Carl budgets 25% of his salary for rent. He makes $250 a week. How much does he budget for rent?
(1) $25   (2) $62.50   (3) $10   (4) $65

7. Paula goes to an English class at night. She got 80% right on a spelling test. There were 25 words on the test. How many words did she spell correctly?
(1) 16   (2) 8   (3) 25   (4) 20

8. Ramon got $12\frac{1}{2}$% raise last week. He was making $160 a week. How much is his raise?
(1) $12.80   (2) $12.50   (3) $19.20   (4) $20

9. In the last election, 40% of the registered voters in Bordenton cast ballots. 11,400 people are registered to vote. How many people voted in the election?
(1) 2,850   (2) 4,660   (3) 285   (4) 4,560

10. The inflation rate for single family housing last year was approximately 5%. A house in Frank's neighborhood cost $66,000 at the first of the year. How much would inflation add to the price by the end of the year?
(1) $3,000   (2) $3,600   (3) $3,300   (4) $4,320

11. This year the number of students at Tri-City Adult High School increased 15%. Last year's enrollment was 840 students. How many students attended Tri-City this year?
(1) 126   (2) 714   (3) 966   (4) 855

12. Due to the lower speed limit, Lumberton County reported an 8% decrease in annual highway fatalities. Last year Lumberton County had 150 highway deaths. How many did they have this year?
(1) 142   (2) 158   (3) 12   (4) 138

13. ATEX stock dropped 6% last week. It was selling for $24 a share at the beginning of the week. What was the price of ATEX stock at the end of the week?
    (1) $22.56   (2) $30   (3) $1.44   (4) $25.44

14. Ron took a cut in pay to take a new job near his house. He was making $16,000 a year on his old job. He took a 4% cut. How much does Ron make on his new job?
    (1) $12,000   (2) $15,630   (3) $15,360   (4) $15,936

15. Mr. Richards bought a boat for $9,400. He paid a 4% sales tax. What was the total cost of the boat?
    (1) $9,776   (2) $9,767   (3) $9,677   (4) $9,800

16. Before Christmas, Carl's Clothes sold men's jackets for $85. After Christmas, the jackets were on sale for 15% off. What was the price of the jackets after Christmas?
    (1) $97.75   (2) $62.25   (3) $72.25   (4) $70

17. Jake owns a hardware store. He pays $16 for an electric drill. Jake charges his customers a 30% markup. How much does a customer pay for the drill?
    (1) $20.20   (2) $13.20   (3) $20.80   (4) $16.30

18. Allen bought a car last year for $3,600. This year the value of the car has depreciated 20%. What is the value of Allen's car this year?
    (1) $3,000   (2) $4,320   (3) $3,460   (4) $2,880

19. David bought a used bicycle for $60. After repairing and repainting it, he wants to make a profit of 35%. For how much will David sell the bicycle?
    (1) $95   (2) $39   (3) $81   (4) $63.50

20. Sylvia wants to buy a new typewriter. The usual price is $249. Tom's Typewriters is selling the model that Sylvia wants for 10% off. What is the price of the typewriter at Tom's?
    (1) $224.10   (2) $225.90   (3) $239   (4) $273.10

Check your answers on page 231.

30. FINDING WHAT PERCENT ONE NUMBER IS OF ANOTHER

Sometimes you will want to find what percent one number is of another. For example, suppose you have $16 and spend $8. What percent of your money have you spent? Ask yourself, "8 is what percent of 16?" or, "What percent of 16 is 8?" To find the answer, write a fraction with the part spent written over the whole amount you had.

Step 1	Step 2	Step 3	Step 4
$\frac{8}{16}$	$\frac{8}{16} = \frac{1}{2}$	$2\overline{)1.0}^{\ \ .5}$	$.5 = 50\%$

Step 1—Compare the two numbers as a fraction. Put the part on top and the whole you had on the bottom.

Step 2—Reduce the fraction if possible.

Step 3—Convert the fraction to a decimal.

Step 4—Convert the decimal to a percent.

Try It (Answers on page 232.)

1. 32 is what percent of 128?

2. What percent of 40 is 30?

3. 2 is what percent of 16?

4. What percent of 90 is 36?

5. 7 is what percent of 70?

6. What percent of 410 is 369?

7. 24 is what percent of 120?

8. _____% of 75 is 15?

9. What percent of 20 is 4?

10. _____% of 25 is 25?

MORE THAN 100%

In some problems the answer may be more than 100%. Look at these examples:

$$24 \text{ is what percent of } 12?$$

$$\frac{24}{12} = 2 = 200\%$$

$$\text{What percent of 10 is 16?}$$

$$\frac{16}{10} = 1\frac{6}{10} = 1.6 = 160\%$$

Try It (Answers on page 232.)

1. 8 is _____% of 4

2. 6 is _____% of 3

3. 12 is _____% of 4

4. 2 is _____% of 1

5. 14 is _____% of 4

6. 6 is _____% of 4

7. 128 is _____% of 32

8. 100 is _____% of 10

9. $1 is what percent of 25¢?

10. 1000 is _____% of 100

PERCENT INCREASE OR DECREASE

Percents can be used to show an increase or decrease. Percent increase or percent decrease problems are worked almost alike.

To find the percent increase, compare the actual increase to the original amount.

$$\text{PERCENT INCREASE} = \frac{\text{AMOUNT OF INCREASE}}{\text{ORIGINAL AMOUNT}}$$

For example, suppose bus fares are now 50¢ and are to go up to 60¢. This is an increase of 10¢ (60 − 50 = 10). To find the percent increase:

$$\frac{\text{AMOUNT OF INCREASE}}{\text{ORIGINAL AMOUNT}} \quad = \quad \frac{10}{50} \quad = \quad \frac{20}{100} \quad = 20\%$$

To find the percent decrease, compare the decrease to the original amount.

$$\boxed{\text{PERCENT DECREASE} \quad = \quad \frac{\text{AMOUNT OF DECREASE}}{\text{ORIGINAL AMOUNT}}}$$

For example, suppose a $25 sweater is on sale for $20. This is a decrease of $5 ($25 − $20 = $5). To find the percent decrease:

$$\frac{\text{AMOUNT OF DECREASE}}{\text{ORIGINAL AMOUNT}} \quad = \quad \frac{5}{25} \quad = \quad 25\overline{)5.0}^{.2} \quad = 20\%$$
$$\underline{5\ 0}$$

Try It (Answers on page 232.)

Find the percent of increase or decrease.

1. From 80° to 100°.

2. From 40° to 20°.

3. From 1,412 ft to 1,059 ft.

4. From $1.00 to 70¢.

5. From 30 mph to 50 mph.

6. From 4 ft to 6 ft.

7. From $400 to $750.

8. From 10¢ to 20¢.

9. From 180 lbs to 162 lbs.

10. From 4 ounces to 16 ounces.

WORKING WITH WORD PROBLEMS

Read the Steps for Solving Word Problems on page xii again. Then work these problems. Find the correct answer.

1. A standard small sedan sold for $4,000 in 1974. Due to inflation, a similar model now sells for $9,000. What is the percent increase in the cost of this type of car?
   (1) 20%   (2) 50%   (3) 90%   (4) 125%

2. The speed limit in the business district of Memphis was reduced from 45 miles per hour to 30 miles per hour. What percent was the speed limit reduced?
   (1) $33\frac{1}{3}$%   (2) $66\frac{2}{3}$%   (3) 50%   (4) 150%

3. Leona got a raise last week. Her salary went from $180 to $192.60 per week. What percent raise did she get?
   (1) 12.6%   (2) 6.5%   (3) 6%   (4) 7%

4. Mr. Leviton's new car depreciated from $8,000 to $7,000 after 6 months. What percent did the car depreciate?
   (1) 23%   (2) 25%   (3) 12.5%   (4) 30%

5. The population of Plainfield was 48,500 in 1986. In 1988, the census showed a population of 63,050. What percent did the population of Plainfield increase?
   (1) $\frac{1}{7}$%   (2) $\frac{1}{8}$%   (3) 125%   (4) 30%

6. Josephine borrowed $5,000 to pay for tuition. One year after she graduated she paid back $1,250. What percent of her loan did Josephine repay?
   (1) 12.5%   (2) 25%   (3) 37.5%   (4) 40%

7. Mr. and Mrs. Garcia own a clothing store. They pay $8.00 for a popular style of shirt. They charge $11.60 for the shirt. What is the percent of markup on the shirt?
   (1) 45%   (2) 36%   (3) 31%   (4) 11.6%

8. Herbert planned to drive 840 miles over the weekend. By Saturday night he had driven 504 miles. What percent of the total distance did Herbert drive by Saturday night?
   (1) 33.6%   (2) 60%   (3) 50.4%   (4) 40%

9. Janet weighed 160 pounds. She went on a diet and lost 30 pounds. What percent of her weight did she lose?

(1) $\frac{3}{16}$ %   (2) $18\frac{3}{4}$ %   (3) 13%   (4) $81\frac{1}{4}$ %

10. The price of a gallon of gasoline went from $1.10 to $1.21. What was the percent of increase in the price of a gallon of gas?

(1) 12%   (2) 11%   (3) 10%   (4) 13%

11. Richard took a math test with 75 problems on it. He got 15 of the problems wrong. What percent of the problems did Richard get right?

(1) 15%   (2) 60%   (3) 75%   (4) 80%

12. In 1979 there were 106 members of the Oakdale Block Association. In 1980 there were 265 members. By what percent did the membership of the association increase?

(1) 159%   (2) 106%   (3) 53%   (4) 150%

13. Manny runs a record shop. On Monday he received 200 copies of a new, popular album. By Friday there were only 24 copies left in the shop. What percent of the records did Manny sell?

(1) 76%   (2) 12%   (3) 98%   (4) 88%

14. There are 25,300 voters in Middletown. At the last election 10,120 people voted in Middletown. What percent of the voters voted in the last election?

(1) 40%   (2) 25.3%   (3) 60%   (4) 11.4%

15. Fred and Ethel bought their house in 1960 for $24,000. They sold their house in 1980 for $39,600. What percent profit did Fred and Ethel make on their house?

(1) 65%   (2) 165%   (3) 61%   (4) 39.6%

Check your answers on page 233.

31. FINDING A NUMBER WHEN A PERCENT OF IT IS GIVEN

You may need to find a number when only a percent of it is given. For example, suppose an article in the newspaper states that 75% of all merchants on Main Street contributed to the Christmas fund. The names of 12 businessmen are then given. How many merchants were asked to contribute? Ask yourself, "12 is 75% of _____?" or, "75% of what number is 12?"

Step 1	Step 2	Step 3
$75\% = \frac{3}{4}$	$12 \div \frac{3}{4} = \frac{\overset{4}{\cancel{12}}}{1} \times \frac{4}{\underset{1}{\cancel{3}}} = 16$	Check: $75\% = \frac{3}{4}$ $\frac{\overset{4}{\cancel{16}}}{1} \times \frac{3}{\underset{1}{\cancel{4}}} = 12$

Step 1—Change the percent to a fraction.

Step 2—Divide the given number by the fraction. Remember to invert the fraction before you multiply. Cancel where possible.

Step 3—To check, multiply the answer by the fraction.

Step 1	Step 2	Step 3
$75\% = .75$	$.75\overline{)12.00}$ gives $16.$ $\underline{7\ 5}$ $4\ 50$ $\underline{4\ 50}$	Check: $\begin{array}{r} 16 \\ \times\ .75 \\ \hline 80 \\ 11\ 2 \\ \hline 12.00 \end{array}$

Step 1—Change the percent to a decimal.

Step 2—Divide the given number by the decimal.

Step 3—To check, multiply the answer by the decimal.

Try It (Answers on page 233.)

Find the missing number.

1. 8 is 25% of _____ .

2. 23 is 50% of _____ .

3. 20 is 20% of _____ .

4. 100 is 40% of _____ .

5. 100% of _____ is 30.

6. 12 $\frac{1}{2}$% of _____ is 2.

7. 90% of _____ is 90.

8. 30% of _____ is 36.

9. 12 is 75% of _____ .

10. 120 is 80% of _____ .

WORKING WITH WORD PROBLEMS

Read the Steps for Solving Word Problems on page xii again. Then work these problems. Find the correct answer.

1. 25% of Laverne's salary goes for rent. She pays $285 a month for her apartment. How much does Laverne make a month?
   (1) $250   (2) $310   (3) $399   (4) $1,140

2. Maria has saved $1200. This is 75% of the down payment she needs for a used car. How much is the total down payment on the car?
   (1) $1600   (2) $1950   (3) $2100   (4) $1450

3. The Forest College Friday game is a 75% sell out. 60,000 tickets have been sold. If all the tickets are sold, how many people will see the game?
   (1) 75,000   (2) 85,000   (3) 80,000   (4) 90,000

4. Lee bought a coat on sale for $48. This was 80% of the original price. What was the original price of the coat?
   (1) $60   (2) $128   (3) $68   (4) $240

5. Telemar Electronics has TVs marked 30% off. The TVs are on sale for $280. What was the original price of the TV?
(1) $310   (2) $350   (3) $340   (4) $400

6. Sal got 34 questions right on his English test. These questions were 85% of the test. Find the total number of questions on the test.
(1) 51   (2) 68   (3) 50   (4) 40

7. 30% of the Miller family's income goes for food. Their average weekly food bill is $61.50. What is their weekly income?
(1) $91.50   (2) $184.50   (3) $205   (4) $123

8. Because of bad weather only 65 % of the members of the local of a labor union attended the last meeting. 117 people attended the meeting. How many members belong to the local?
(1) 180   (2) 76   (3) 130   (4) 234

9. Mike has saved $825 in order to buy a motorcycle. This is 60% of the amount that he needs. How much does Mike need altogether for the motorcycle?
(1) $885   (2) $4,950   (3) $1,375   (4) $1,650

10. Sam made a long weekend trip. The first day he drove 704 miles. This was 55% of the total distance he planned to drive. Find the total distance of Sam's trip.
(1) 1408 mi   (2) 3872 mi   (3) 1564 mi   (4) 1280 mi

11. Christine now weighs 112 pounds. This is 80% of what she weighed before her diet. How much did Christine weigh before her diet?
(1) 192 lb   (2) 140 lb   (3) 132 lb   (4) 120 lb

12. Al bought a stereo on sale for $144.95. The sale price was 65% of the original price. What was the original price of the stereo?
(1) $200   (2) $209.95   (3) $223   (4) $240

13. Mr. and Mrs. James have paid $11,664 toward the price of their house. This amount represents 36% of the total they must pay. Find the total price of the house.
(1) $41,990   (2) $34,200   (3) $24,300   (4) $32,400

14. Joyce has earned 96 credits. These are 75% of the total credits she needs to graduate. How many credits does Joyce need in order to graduate?
    (1) 100   (2) 171   (3) 128   (4) 124

15. On Friday night 328 people attended the Midtown Theatre. They are 82% of the capacity of the theatre. What is the seating capacity of the Midtown Theatre?
    (1) 269   (2) 400   (3) 375   (4) 410

Check your answers on page 234.

32. SIMPLE INTEREST

When money is borrowed, you must pay to use it. What you pay to use money is called <u>interest</u>. The rate of interest is a percent. The money you borrow is called the <u>principal</u>. You borrow money for a period of time. At the end of the time, you pay back the principal plus the interest. This is called simple interest.

You can receive interest if you put your money in a savings account or other type of investment. In this case, the principal is the money you put in the bank. The interest is the money the bank pays you for using your money.

THE INTEREST FORMULA

You can quickly find the amount of simple interest on a loan. Multiply the principal by the rate of interest *and* the amount of time the money was borrowed for.

$$\underline{\text{I}}\text{nterest} = \underline{\text{P}}\text{rincipal} \times \underline{\text{R}}\text{ate} \times \underline{\text{T}}\text{ime}$$

or simply

$$\text{I} = \text{P R T}$$

I = interest   P = principal   R = rate of interest   T = time

This is called a <u>formula</u>. A formula uses letters to stand for words.

RATE OF INTEREST

The rate of interest is always given as a percent. For example, some government bonds give you a 6% annual interest rate. To work interest problems, you must express the percent as a fraction or as a decimal. Review pages 38 and 40 to see how to change percents to fractions and decimals. Then work this review exercise.

Try It (Answers on page 235.)

Change these percents to fractions and reduce if possible.

1. 10%   2. $33\frac{1}{3}$%   3. 25%   4. $12\frac{1}{2}$%   5. 5%

Change these percents to decimals.

6. 6%   7. 12%   8. 15%   9. 48%   10. $1\frac{1}{2}$%

TIME

Time in the formula is always expressed in years or parts of a year. Look at these examples:

$$4 \text{ months} = \frac{4}{12} = \frac{1}{3} \qquad 5 \text{ years} = \frac{5}{1} \qquad 3\frac{1}{2} \text{ years} = \frac{7}{2}$$

Try It (Answers on page 235.)

Change these periods of time to years or parts of a year. Some answers will be improper fractions. Remember that improper fractions have a top number that is larger than the bottom number.

1. 6 months   2. 2 years   3. 3 months   4. 9 months

5. 1 year   6. 1 year, 6 months   7. 15 months

8. $2\frac{1}{2}$ years   9. 1 month   10. 1 year, 3 months

CALCULATING INTEREST

To find simple interest, use the formula I = PRT. Look at the example on the next page. It is worked two ways, using fractions and using decimals. Either way is correct.

Rod borrowed $500 at 6% interest for 6 months. How much must he pay back at the end of 6 months?

P = amount of principal borrowed = $500

R = rate of interest = 6 % = $\frac{6}{100}$ = $\frac{3}{50}$ OR .06

T = time = 6 months = $\frac{6}{12}$ = $\frac{1}{2}$ OR $12\overline{)6.0}$ = .5

Using fractions:

Step 1	Step 2	Step 3
I = PRT	R $= \frac{6}{100} = \frac{3}{50}$ T $= \frac{6}{12} = \frac{1}{2}$	$\frac{\overset{10}{\cancel{500}}}{1} \times \frac{3}{\underset{1}{\cancel{50}}} \times \frac{1}{\underset{1}{\cancel{2}}} = 15$

Step 1—Write the formula: Interest = principal × rate × time.

Step 2—Change the rate, given as a percent, to a fraction and reduce. Set up time as a fraction of a year.

Step 3—Multiply principal times rate times time. Cancel where possible. 50 and 500 can both be divided by 50. The 10 you get by canceling and the 2 can both be divided by 2.

Using decimals:

Step 1	Step 2	Step 3
I = PRT	$ 5 00 × .06 $30.00	$ 3 0 × .5 $15.0

Step 1—Write the formula: Interest = principal × rate × time.

Step 2—Change the rate, given as a percent, to a decimal (6% = .06) and multiply times the principal.

Step 3—Change the fraction for time to a decimal and multiply times the amount in Step 2.

The total amount repaid on this loan is the $500 borrowed, plus the interest of $15: $500 + $15 = $515.

WORKING WITH WORD PROBLEMS

Read the Steps for Solving Word Problems on page xii again. Then work these problems. Find the correct answer.

1. What is the interest on $800 for 2 years at 8%?
   (1) $128   (2) $16   (3) $256   (4) $12.80

2. What is the interest on $1,400 for 1 year 6 months at 9%?
   (1) $189   (2) $198   (3) $168   (4) $169

3. Vivian bought a $100 bond that pays 6% interest. How much interest will she get in 1 year?
   (1) $.60   (2) $60   (3) $106   (4) $6

4. Mr. Garcia borrowed $2,000 from his bank to buy a used car. He signed a note to repay the loan in 1 year. The bank charged Mr. Garcia 8% interest. How much will Mr. Garcia pay in interest?
   (1) $8   (2) $80   (3) $160   (4) $1,600

5. Marlo borrowed $1,550 from a friend to buy new living room furniture. She was charged 6% interest. She repaid the money plus interest in 6 months. How much interest did she pay her friend?
   (1) $93.30   (2) $25.82   (3) $45.60   (4) $46.50

6. Mr. Keene borrowed $4,000 at 12% to expand his business. He repaid the loan in 2 years, 6 months. How much interest did he pay?
   (1) $12.00   (2) $1,200   (3) $120   (4) $5,200

7. Delia lent a friend $1,200 at 6% for 2 years, 3 months. How much interest will she get on this loan?
   (1) $261   (2) $162   (3) $432   (4) $62

8. Celia and Dan borrowed $1,600 from the Credit Union for 6 months at 8%. How much will they pay back?
   (1) $46   (2) $64   (3) $1,646   (4) $1,664

9. Radmer Corporation borrowed $250,000 at 12% for 9 months. How much did Radmer repay when the loan came due?
   (1) $22,500   (2) $262,000   (3) $270,000   (4) $272,500

10. Mr. and Mrs. Blackmer signed a note for 3 months. The note was to repay $1,000 they borrowed for medical expenses. They were charged 8% interest. How much did they repay when the note came due?
    (1) $1,020   (2) $1,024   (3) $1,022.50   (4) $22.50

Check your answers on page 235.

33. MEASURING YOUR PROGRESS

TEST A

Fill in the blanks.

1. 45 is _____% of 180

2. 90% of 60 = _____

3. 30% of _____ is 21

4. 22 is _____% of 176

5. 48 is _____% of 64

6. $16\frac{2}{3}$% of 90 = _____

7. 192 is 60% of _____

8. 35 is _____% of 56

9. 45% of 80 = _____

10. 72 is 45% of _____

11. 125% of 64 = _____

12. 21 is _____% of 140

13. 24 is _____% of 60

14. 87% of 400 = _____

15. 20% of 40 = _____

16. $33\frac{1}{3}$% of 72 = _____

17. 80 is _____% of 20

18. 16 is 80% of _____

19. 75 is 25% of _____

20. 68 is _____% of 85

222

TEST B

Read the Steps for Solving Word Problems on page xii again. Then work these problems. Find the correct answer.

1. A $488 TV is advertised in the newspaper at 15% off for next week's sale. What will be the price of the TV on sale?
   (1) $14.80   (2) $473   (3) $418.40   (4) $414.80

2. South State Loan lent Gino $792 for a year and 9 months. They charged him 9% interest. How much will he repay?
   (1) $961.74   (2) $873   (3) $963   (4) $916.74

3. By paying cash, Phillip got a stereo for his new apartment for $720. The set sold regularly for $800. What percent did he save?
   (1) 8%   (2) 10%   (3) 7.2%   (4) 80%

4. Hazel has been dieting. She now weighs 135 pounds. This is 90% of her weight last Christmas. How much did she weigh at Christmas?
   (1) 145 lbs   (2) 150 lbs   (3) $148\frac{1}{2}$ lbs   (4) 144 lbs

5. Bill paid 15% more this year for rent than he did last year. Last year he paid $2,280 for rent. How much rent did he pay this year?
   (1) $342   (2) $150   (3) $2,430   (4) $2,622

6. Real Sell Real Estate charges a commission of 7% on the sale of houses. Real Sell sold Mr. and Mrs. North's house for $42,000. After Real Sell took their commission, how much did the Norths get from the sale? (This is called *net proceeds*.)
   (1) $44,940   (2) $41,300   (3) $39,600   (4) $39,060

7. American InterTel expects 25% more losses this year than last. Last year the company lost $84,000. How much do they expect to lose this year?
   (1) $21,000   (2) $105,000   (3) $63,000   (4) $107,000

8. Good Buy Used Cars adds 40% to car prices for their profit. If they buy a used car for $4,200, what will be their asking price for the car?
   (1) $4,600   (2) $5,840   (3) $5,880   (4) $5,800

9. Pablo earns $250 a week selling window awnings. He is paid on commission. He gets 20% commission on everything he sells. How much does Pablo have to sell each week to earn $250?
   (1) $1,250   (2) $2,000   (3) $500   (4) $1,500

10. Carla Simpson bought a car at a year end clearance for $5,600. The car was reduced $12\frac{1}{2}$% from the original sticker price. What was the original price?
    (1) $6,850   (2) $6,400   (3) $8,650   (4) $5,612.50

11. Phil bought a camera for $124. He sold it one year later at a 30% loss. For how much did Phil sell his camera?
    (1) $94   (2) $88.60   (3) $161.20   (4) $86.80

12. On a math test with 40 problems, Alex got 12 problems wrong. What percent of the problems did he get right?
    (1) 30%   (2) 70%   (3) 88%   (4) 60%

13. Susan left $1,200 in her savings account at 5% interest for nine months. How much did she have in the account at the end of nine months?
    (1) $1,205   (2) $1,230   (3) $1,245   (4) $1,260

14. Jack has paid $2,700 on a loan. This amount is 60% of the total loan. Find the total amount that Jack borrowed.
    (1) $4,500   (2) $4,300   (3) $4,100   (4) $3,300

15. Fred borrowed $5,000 at 12% interest for 2 years and 6 months. How much interest did Fred pay on his loan?
    (1) $600   (2) $900   (3) $1,200   (4) $1,500

16. In 1970 the population of Central City was 24,000. In 1980 the population was 28,800. By what percent did the population of Central City increase?
    (1) $16\frac{2}{3}$%   (2) 24%   (3) 48%   (4) 20%

17. At last Thursday's meeting of the Kendall Projects Tenants' Association 156 people were present. They are 65% of the total membership. How many people belong to the association?
(1) 1,014   (2) 200   (3) 240   (4) 221

18. Franz pays $275 a month for rent. His rent is 25% of his monthly income. What is Franz's monthly income?
(1) $1,100   (2) $900   (3) $800   (4) $700

19. Janice hoped to drive 400 miles to her parents' home before sundown. However, at sundown she had driven only 304 miles. What percent of the distance did she have left to drive?
(1) 24%   (2) 48%   (3) 72%   (4) 96%

20. Joan bought a rocking chair for $115. The tax in her state is 8%. What was the cost of the chair including tax?
(1) $124.20   (2) $123   (3) $107   (4) $105.80

21. In January Maureen weighed 152 pounds. By June she weighed 133 pounds. What percent of her weight did she lose?
(1) 19%   (2) 14%   (3) $12\frac{1}{2}\%$ (4) 9%

22. The Uptown Community Organization has collected $2,624 to install new street lights in their neighborhood. The money is 80% of the total they need. Find the total amount the organization needs for the project.
(1) $3,280   (2) $3,424   (3) $2,704   (4) $3,820

23. Mr. Johnson manages a grocery store. He pays 50¢ for a quart of milk. He charges his customers 56¢ for the milk. Find the percent of markup that Mr. Johnson charges.
(1) 6%   (2) 12%   (3) 11%   (4) 8%

24. When Jackie started night school there were 30 people in her class. By the end of the term 20% of the students had dropped out. How many students were in the class at the end of the term?
(1) 6   (2) 10   (3) 24   (4) 28

25. Sally works in an office 260 days in a year. Last year she was absent from work 13 times. What percent of the time was Sally absent last year?
(1) 26%   (2) 13%   (3) 10%   (4) 5%

Check your answers on page 236.

ANSWERS

HOW MUCH DO YOU KNOW ALREADY?, PAGE 190

Put an X by each problem you missed. Put an X by each problem you did not do. The problems are listed under the titles of the chapters in this book. If you have <u>one or more Xs</u> in a section, study that chapter to improve your skills.

WHAT IS A PERCENT? (pages 193–200)

1.

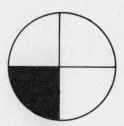

2.

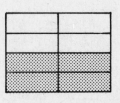

3.

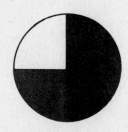

4. **.06**   5. **1.23**   6. **.125**

7. **3%**   8. **76%**   9. **.7%**

10. $\frac{7}{100}$ 11. $\frac{9}{100}$ 12. $\frac{1}{5}$

13. $\frac{3}{4}$ 14. $\frac{3}{400}$ 15. $\frac{1}{25}$

FINDING A PERCENT OF A NUMBER (pages 201–207)

16. **2**   17. **12**   18. **31**

19. **150**   20. **104**   21. **91**

FINDING WHAT PERCENT ONE NUMBER IS
OF ANOTHER NUMBER (pages 208–212)

22. **25%**   23. **20%**   24. **250%**

FINDING A NUMBER WHEN A PERCENT OF IT IS GIVEN
(pages 213–216)

25. **32**   26. **3**   27. **16**

SIMPLE INTEREST (pages 217–221)

28. **$48**   29. **$824**

TRY IT, PAGE 194

1. **10%**	2. **25%**	3. **40%**	4. **30%**
90%	**75%**	**60%**	**70%**
100%	**100%**	**100%**	**100%**
5. **75%**	6. **15%**	7. **50%**	8. **60%**
25%	**85%**	**50%**	**40%**
100%	**100%**	**100%**	**100%**

TRY IT, PAGE 195

1. **.15**	2. **.37**	3. **.63**	4. **.09**	5. **.67**
6. **.10 or .1**	7. **.125**	8. **.06**	9. **.03**	10. **.99**
11. **1**	12. **3.5**	13. **1.75**	14. **2.5**	15. **3**

TRY IT, PAGE 196, TOP

1. **3%**	2. **20%**	3. **75%**	4. **60%**	5. **19%**
6. **.7%**	7. **7.5%**	8. **8.25%**	9. **90%**	10. **67%**

TRY IT, PAGE 196, BOTTOM

1. **.0075**	2. **.0010 or .001**	3. **.002**	4. **.003**	5. **.009**
6. **.005**	7. **.0025**	8. **.008**	9. **.006**	10. **.004**

TRY IT, PAGE 197

1. $\frac{9}{100}$ 2. $\frac{2}{25}$ 3. $\frac{1}{5}$ 4. $\frac{7}{20}$ 5. $\frac{43}{50}$

6. $\frac{3}{4}$ 7. $\frac{2}{5}$ 8. $\frac{1}{20}$ 9. $\frac{4}{5}$ 10. $\frac{3}{50}$

TRY IT, PAGE 198

1. **20**%   2. **50**%   3. **12.5**%   4. **5**%   5. **25**%

6. **37.5**%   7. **33$\frac{1}{3}$**%   8. **40**%   9. **87.5**%   10. **66$\frac{2}{3}$**%

TRY IT, PAGE 199

1. 150%   **1$\frac{1}{2}$** or $\frac{3}{2}$ **1.5**

2. **25**% $\frac{1}{4}$ .25

3. **50**% $\frac{1}{2}$ **.5**

4. 75% $\frac{3}{4}$ **.75**

5. **10**% $\frac{1}{10}$ **.1**

6. **20**% $\frac{1}{5}$ .2

7. 30% $\frac{3}{10}$ **.3**

8. **35**% $\frac{7}{20}$ .35

9. **65**% $\frac{13}{20}$ **.65**

10. 62.5% $\frac{5}{8}$ **.625**

WORKING WITH WORD PROBLEMS, PAGE 199

1. **(3)** $\frac{50}{100} = $ **50**%   2. **(4)** $\frac{3}{10} = \frac{30}{100} = $ **30**%

3. **(1)** $\frac{65}{100} = $ **65**%   4. **(3)**   .347 = **34.7**%

5. **(2)** $\frac{2}{10} = \frac{20}{100} = $ **20**%   6. **(3)**   75% $= \frac{75}{100} = \frac{3}{4}$

7. **(1)**   20% = .20

$$100\% = \quad 1.00$$
$$- \quad .20$$
$$.80$$

$$.80 = \frac{80}{100} = \frac{4}{5}$$

8. **(1)** $\frac{9}{20} = 20\overline{)9.00}$ gives $.45$

$$20\overline{)9.00}$$
$$\underline{8\ 0}$$
$$1\ 00$$
$$\underline{1\ 00}$$

$$1.00$$
$$- \quad .45$$
$$.55$$

$$.55 = \mathbf{55\%}$$

9. **(2)** $\frac{1}{16} = 16\overline{)1.0000}$ gives $.0625$

$$16\overline{)1.0000}$$
$$\underline{96}$$
$$40$$
$$\underline{32}$$
$$80$$
$$\underline{80}$$

$$.0625 = \mathbf{6.25\%}$$

10. **(3)** $5 \times \frac{1}{8} = \frac{5}{8}$

$$\frac{5}{8} = 8\overline{)5.0} \text{ gives } .625$$
$$\underline{4\ 8}$$
$$20$$
$$\underline{16}$$
$$40$$
$$\underline{40}$$

$$100\% = \quad 1.00$$
$$- \quad .625$$
$$.375$$

$$.375 = \quad \mathbf{37.5\%}$$

TRY IT, PAGE 202

1. **48**	2. **64**	3. **24**	4. **64**
5. **60**	6. **28**	7. **450**	8. **12**
9. **16**	10. **20**	11. **2**	12. **1.5**

TRY IT, PAGE 203

1. $\frac{3}{4}$	**150**	**148**
2. $\frac{1}{10}$	**30**	**27**
3. $\frac{1}{4}$	**24**	**23.04**
4. $\frac{1}{2}$	**43**	**42.14**

5. $\frac{1}{3}$ **3**   **2.88**

WORKING WITH WORD PROBLEMS, PAGE 205

1. **(2)**   8 29
        $\times$ .10
      $82.90

2. **(4)** $\frac{1}{\cancel{7}} \times \frac{\overset{21}{\cancel{84}}}{1} =$ **$21**

3. **(3)**   56
        $\times$ .06
      $3.36

4. **(1)** $\frac{1}{\cancel{8}} \times \frac{\overset{\$1600}{\cancel{8000}}}{1} =$ **$1600**

5. **(2)** $\frac{1}{\cancel{3}} \times \frac{\overset{3000}{\cancel{9000}}}{1} =$ **3000**

6. **(2)** $\frac{1}{\cancel{4}} \times \frac{\overset{\$62.50}{\cancel{250}}}{1} =$ **$62.50**

7. **(4)**   25
        $\times$ .80
      20.00

8. **(4)** $\frac{1}{\cancel{8}} \times \frac{\overset{20}{\cancel{160}}}{1} =$ **$20**

9. **(4)**   114 00
        $\times$ .40
      4,560.00

10. **(3)**   66000
        $\times$ .05
      $3,300

11. **(3)**   840   840
        $\times$ .15   + 126
        126   **966**

12. **(4)**   1 50   150   1 50
        $\times$ .08 $-$ 12   **or** $\times$ 1.08
        12.00   **138**   **1 38**

13. **(1)** $24 $24.00
        $\times$ .06 $-$ 1.44
        $1.44   **$22.56**

14. **(3)** $160 00 $16 000
        $\times$ .04 $-$ 640
        $6.40   **$15,360**

15. **(1)** $9400 $9 400
        $\times$ .04   +   376
        $376   **$9,776**

16. **(3)** $85 $85.00
        $\times$ .15 $-$ 12.75
        4 25   **$72.25**
        8 5
        $12.75

17. **(3)** $16 $16.00
        $\times$ .30   + 4.80
        $4.80   **$20.80**

18. **(4)** $36 00 $3600
        $\times$ .20 $-$ 720
        $720.00   **$2880**

19. **(3)**
$$\begin{array}{r} \$60 \\ \times \quad .35 \\ \hline 3\ 00 \\ 18\ 0 \\ \hline \$21.00 \end{array}$$

$$\begin{array}{r} \$60 \\ + \quad 21 \\ \hline \mathbf{\$81} \end{array}$$

20. **(1)**
$$\begin{array}{r} 2\ 49 \\ \times \quad .10 \\ \hline \$24.90 \end{array}$$

$$\begin{array}{r} \$249.00 \\ - \quad 24.90 \\ \hline \mathbf{\$224.10} \end{array}$$

TRY IT, PAGE 208

1. **25%**	2. **75%**
3. **12½% or 12.5%**	4. **40%**
5. **10%**	6. **90%**
7. **20%**	8. **20%**
9. **20%**	10. **100%**

TRY IT, PAGE 209

1. **200%**	2. **200%**
3. **300%**	4. **200%**
5. **350%**	6. **150%**
7. **400%**	8. **1000%**
9. **400%**	10. **1000%**

TRY IT, PAGE 210

1. **25%**	2. **50%**
3. **25%**	4. **30%**
5. **66⅔%**	6. **50%**
7. **87½% or 87.5%**	8. **100%**
9. **10%**	10. **300%**

WORKING WITH WORD PROBLEMS, PAGE 211

1. **(4)**   9,000 $\frac{5,000}{4,000} = \frac{5}{4} =$ **125%**
     − 4,000
     5,000

2. **(1)**   45 $\frac{15}{45} = \frac{1}{3} =$ **33 $\frac{1}{3}$%**
     − 30
     15

3. **(4)**   192.60 $\frac{12.60}{180} = .07 =$ **7%**
     − 180.00
     12.60

4. **(3)**   8,000 $\frac{1,000}{8,000} = \frac{1}{8} =$ **12 $\frac{1}{2}$%** or **12.5%**
     − 7,000
     1,000

5. **(4)**   63,050 $\frac{14,550}{48,500} = .3 =$ **30%**
     − 48,500
     14,550

6. **(2)** $\frac{1250}{5000} = \frac{1}{4} =$ **25%**

7. **(1)**   11.60 $\frac{3.60}{8.00} = \frac{9}{20} =$ **45%**
     − 8.00
     3.60

8. **(2)** $\frac{504}{840} = \frac{3}{5} =$ **60%**

9. **(2)** $\frac{30}{160} = \frac{3}{16} =$ **18 $\frac{3}{4}$%**

10. **(3)**   1.21 $\frac{.11}{1.10} = \frac{1}{10} =$ **10%**
     − 1.10
     .11

11. **(4)**   75 $\frac{60}{75} = \frac{4}{5} =$ **80%**
     − 15
     60

12. **(4)**   265 $\frac{159}{106} = \frac{3}{2} =$ **150%**
     − 106
     159

13. **(4)**   200 $\frac{176}{200} =$ **88%**
     − 24
     176

14. **(1)** $\frac{10,120}{25,300} = \frac{2}{5} =$ **40%**

15. **(1)**   39,600· $\frac{15,600}{24,000} = \frac{13}{20} =$ **65%**
     − 24,000
     15,600

TRY IT, PAGE 214

1. **32**

2. **46**

3. **100**

4. **250**

5. **30**

6. **16**

7. **100**

8. **120**

9. **16**

10. **150**

WORKING WITH WORD PROBLEMS, PAGE 214

1. **(4)** $285 \div \frac{1}{4} = \frac{285}{1} \times \frac{4}{1} =$ **$1,140**   2. **(1)**

$$\begin{array}{r} \$16\ 00. \\ .75\overline{)1200.00} \\ 75 \\ \hline 450 \\ 450 \\ \hline \end{array}$$

or $1200 \div \frac{3}{4} = \frac{\overset{400}{\cancel{1200}}}{1} \times \frac{4}{\underset{1}{\cancel{3}}} =$ **$1600**

3. **(3)** $60,000 \div \frac{3}{4} = \frac{\overset{20000}{\cancel{60000}}}{1} \times \frac{4}{\underset{1}{\cancel{3}}} =$ **80,000**   4. **(1)**

$$\begin{array}{r} \$60. \\ .80\overline{)48.00.} \end{array}$$

5. **(4)** 30% off means the set sold for 70% of its original price.

100% − 30% = 70%

$$\begin{array}{r} \$4\ 00 \\ .70\overline{)280.00} \\ 280 \\ \hline \end{array}$$

6. **(4)**
$$\begin{array}{r} 40 \\ .85\overline{)34.00} \end{array}$$

7. **(3)**
$$\begin{array}{r} \$20\ 5 \\ .3\overline{)\ 61.5} \end{array}$$

8. **(1)**
$$\begin{array}{r} 1\ 80 \\ .65\overline{)117.00} \end{array}$$

9. **(3)**
$$\begin{array}{r} \$137\ 5 \\ .6\overline{)\ 825.0} \end{array}$$

10. **(4)**
$$\begin{array}{r} 12\ 80 \\ .55\overline{)704.00} \end{array}$$

11. **(2)**
$$\begin{array}{r} 14\ 0 \\ .8\overline{)112.0} \end{array}$$

12. **(3)**
$$\begin{array}{r} \$2\ 23 \\ .65\overline{)144.95} \end{array}$$

13. **(4)**
$$\begin{array}{r} \$32,4\ 00 \\ .36\overline{)11,66\ 4.00} \end{array}$$

14. **(3)**
$$\begin{array}{r} 1\ 28 \\ .75\overline{)96.00} \end{array}$$

15. **(2)**
$$\begin{array}{r} 4\ 00 \\ .82\overline{)328.00} \end{array}$$

TRY IT, PAGE 218, TOP

1. $\frac{1}{10}$ 2. $\frac{1}{3}$ 3. $\frac{1}{4}$ 4. $\frac{1}{8}$ 5. $\frac{1}{20}$

6. **.06** 7. **.12** 8. **.15** 9. **.48** 10. **.015**

TRY IT, PAGE 218, BOTTOM

1. $\frac{1}{2}$ 2. **2** 3. $\frac{1}{4}$ 4. $\frac{3}{4}$ 5. **1**

6. $\frac{3}{2}$ 7. $\frac{5}{4}$ 8. $\frac{5}{2}$ 9. $\frac{1}{12}$ 10. $\frac{5}{4}$

WORKING WITH WORD PROBLEMS, PAGE 220

1. **(1)** $\dfrac{\overset{8}{\cancel{800}}}{1} \times \dfrac{8}{\cancel{100}} \times \dfrac{2}{1} = $ **$128**

2. **(1)** $\dfrac{\overset{7}{\cancel{\overset{14}{\cancel{1400}}}}}{1} \times \dfrac{9}{\cancel{100}} \times \dfrac{3}{2} = $ **$189**

3. **(4)** $\dfrac{\overset{1}{\cancel{100}}}{1} \times \dfrac{6}{\cancel{100}} = $ **$6**

4. **(3)** $\dfrac{\overset{20}{\cancel{2000}}}{1} \times \dfrac{8}{\cancel{100}} = $ **$160**

5. **(4)** $\dfrac{\overset{31}{\cancel{1550}}}{1} \times \dfrac{3}{\cancel{50}} \times \dfrac{1}{2} = \dfrac{93}{2} = $ **$46.50**

6. **(2)** $\dfrac{\overset{20}{\cancel{\overset{40}{\cancel{4000}}}}}{1} \times \dfrac{12}{\cancel{100}} \times \dfrac{5}{2} = $ **$1,200**

7. **(2)** $\dfrac{\overset{3}{\cancel{1200}}}{1} \times \dfrac{6}{\cancel{100}} \times \dfrac{9}{\cancel{4}} = $ **$162**

8. **(4)** $\dfrac{\overset{16}{\cancel{1600}}}{1} \times \dfrac{8}{\cancel{100}} \times \dfrac{1}{2} = 64$

$$\begin{array}{r} 1\,600 \\ +\quad 64 \\ \hline \mathbf{\$1,664} \end{array}$$

9. **(4)** $\dfrac{\overset{2500}{\cancel{250,000}}}{1} \times \dfrac{3}{\cancel{12}} \times \dfrac{3}{4} = 22,500$

$$\begin{array}{r} 250,000 \\ +\quad 22,500 \\ \hline \mathbf{\$272,500} \end{array}$$

10. **(1)** $\dfrac{\overset{10}{\cancel{1000}}}{1} \times \dfrac{2}{\cancel{100}} \times \dfrac{1}{4} = 20$

$$\begin{array}{r} 1\,000 \\ +\quad 20 \\ \hline \mathbf{\$1,020} \end{array}$$

MEASURING YOUR PROGRESS, PAGE 222

Test A

1. **25**	2. **54**	3. **70**
4. **12½**	5. **75**	6. **15**
7. **320**	8. **62½**	9. **36**
10. **160**	11. **80**	12. **15**
13. **40**	14. **348**	15. **8**
16. **24**	17. **400**	18. **20**
19. **300**	20. **80**	

Test B

1. **(4)**
$$\begin{array}{r} 4.88 \\ \times\ \ .15 \\ \hline 73.20 \end{array}$$
$$\begin{array}{r} 488.00 \\ -\ \ 73.20 \\ \hline \mathbf{\$414.80} \end{array}$$

2. **(4)** $792 \times .09 \times 1.75 = 124.74$
$$\begin{array}{r} 792.00 \\ +\ 124.74 \\ \hline \mathbf{\$916.74} \end{array}$$

or

$100 - 15 = 85$
$$\begin{array}{r} 4.88 \\ \times\ \ .85 \\ \hline \mathbf{\$414.80} \end{array}$$

3. **(2)**
$$\begin{array}{r} 800 \\ -\ 720 \\ \hline 80 \end{array}$$
$\frac{80}{800} = \frac{1}{10} = \mathbf{10\%}$

4. **(2)**
$$\begin{array}{r} 1\ 50. \\ .90\overline{)135.00.} \\ \underline{90} \\ 45\ 0 \\ \underline{45\ 0} \end{array}$$

5. **(4)**
$$\begin{array}{r} 2,280 \\ \times\ \ .15 \\ \hline 342 \end{array}$$
$$\begin{array}{r} 2,280 \\ +\ \ \ 342 \\ \hline \mathbf{\$2,622} \end{array}$$

6. **(4)**
$$\begin{array}{r} 42,000 \\ \times\ \ \ \ .07 \\ \hline 2,940 \end{array}$$
$$\begin{array}{r} 42,000 \\ -\ \ 2,940 \\ \hline \mathbf{\$39,060} \end{array}$$

7. **(2)** $25\% = \frac{1}{4}$
$\frac{\overset{21000}{\cancel{84000}}}{1} \times \frac{1}{\underset{1}{\cancel{4}}} = 21,000$
$$\begin{array}{r} 84\ 000 \\ -\ 21\ 000 \\ \hline \mathbf{\$105,000} \end{array}$$

8. **(3)**
$$\begin{array}{r} 4,200 \\ \times\ \ \ .40 \\ \hline 1,680 \end{array}$$
$$\begin{array}{r} 4,200 \\ +\ 1,680 \\ \hline \mathbf{\$5,880} \end{array}$$

9. **(1)** $20\% = \frac{1}{5}$ $250 \div \frac{1}{5} = \frac{250}{1} \times \frac{5}{1} = $ **$1250**

10. **(2)** $12\frac{1}{2}\% = \frac{1}{8}$ OFF   PAID $\frac{7}{8}$

$5{,}600 \div \frac{7}{8} = \frac{\cancel{5{,}600}^{800}}{1} \times \frac{8}{\cancel{7}} = $ **$6,400**

11. **(4)**
$$\begin{array}{r} \$1\,24 \\ \times \quad .30 \\ \hline \$37.20 \end{array} \qquad \begin{array}{r} \$124.00 \\ - \quad 37.20 \\ \hline \mathbf{\$86.80} \end{array}$$

12. **(2)**
$$\begin{array}{r} 40 \\ -\,12 \\ \hline 28 \end{array} \qquad \frac{28}{40} = \frac{7}{10} = \mathbf{70\%}$$

13. **(3)** $\dfrac{\cancel{\cancel{1200}}^{3}}{1} \times \dfrac{5}{\cancel{100}} \times \dfrac{3}{\cancel{4}} = \$45$

$$\begin{array}{r} \$1200 \\ + \quad 45 \\ \hline \mathbf{\$1245} \end{array}$$

14. **(1)** $.6)\overline{2700.0} \quad \dfrac{\$450\ 0}{\ }$

15. **(4)** $\dfrac{\cancel{5000}^{50}}{1} \times \dfrac{\cancel{12}^{6}}{\cancel{100}} \times \dfrac{5}{\cancel{2}} = $ **$1500**

16. **(4)**
$$\begin{array}{r} 28{,}800 \\ -\,24{,}000 \\ \hline 4{,}800 \end{array} \qquad \frac{4{,}800}{24{,}000} = \frac{1}{5} = \mathbf{20\%}$$

17. **(3)** $.65)\overline{156.00} \quad \dfrac{2\ 40}{\ }$

18. **(1)** $.25)\overline{\$275.00} \quad \dfrac{\$11\ 00}{\ }$

19. **(1)**
$$\begin{array}{r} 400 \\ -\,304 \\ \hline 96 \end{array} \qquad \frac{96}{400} = \frac{24}{100} = \mathbf{24\%}$$

20. **(1)**

$$\begin{array}{r} \$1\ 15 \\ \times\ \ \ .08 \\ \hline \$9.20 \end{array}$$

$$\begin{array}{r} \$115.00 \\ +\ \ \ \ 9.20 \\ \hline \mathbf{\$124.20} \end{array}$$

21. **(3)**

$$\begin{array}{r} 152 \\ -\ 133 \\ \hline 19 \end{array}$$

$\frac{19}{152} = \frac{1}{8} = \mathbf{12\frac{1}{2}\%}$

22. **(1)**

$$\overset{\mathbf{\$328\ 0}}{.8)\overline{\$2624.0}}$$

23. **(2)**

$$\begin{array}{r} 56 \\ -\ 50 \\ \hline 6 \end{array}$$

$\frac{6}{50} = \mathbf{12\%}$

24. **(3)**

$$\begin{array}{r} 30 \\ \times\ \ .20 \\ \hline 6.00 \end{array}$$

$$\begin{array}{r} 30 \\ -\ \ \ 6 \\ \hline \mathbf{24} \end{array}$$

25. **(4)** $\frac{13}{260} = \frac{1}{20} = \mathbf{5\%}$

Charts and Graphs

34. HOW MUCH DO YOU KNOW ALREADY?

Here are some problems with charts and graphs. You will be asked to read and interpret them. This will help you find out how much you know already. You will also find out what you need to study. Do as many problems as you can. Write your answers on the lines at the right. Then, turn to page 268. Check your answers. Page 268 will tell you what pages in the book to study.

1. How many work accidents occurred in 1990?

2. What percent of accidents were caused by burns and cuts?

1. _____

2. _____

**Accidental Injuries
Springfield**

YEAR	AUTO	HOME	WORK
'80	23	72	35
'85	38	97	52
'90	42	129	43
'95	40	150	30

**Causes of
Accidents**

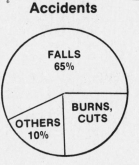

3. There are 250 students at night school. How many took science?

4. How many students graduated from Roosevelt?

Night School Classes

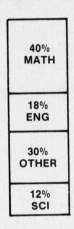

3. _____

4. _____

Local Graduates

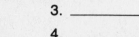

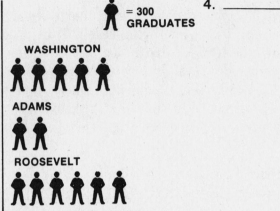

5. What is the elevation of Billings Mountain?

6. Between which two months did rainfall decrease the most?

Mountain Elevations

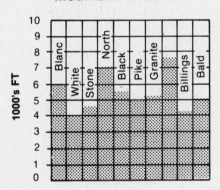

Springfield 1989 Rainfall

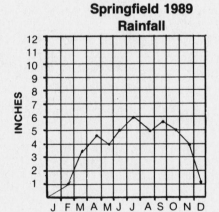

5. _____

6. _____

35. CHARTS

A chart or table is a listing of data. It is a concise way to show numerical data. Tables or charts consist of columns and rows of data. Look at the example below:

NFL Passing Records

YEAR	ATTEMPTS	COMPLETIONS	YARDS GAINED	TOUCHDOWNS
1968	317	182	2909	17
1969	422	274	3102	15
1970	378	223	2941	24
1971	211	126	1882	15
1972	325	196	2307	17

In 1971, how many yards were gained by the NFL's leading passer? Find 1971 under the "year" column. Read across the 1971 row to the column marked "yards gained." Find 1882. The passing record in 1971 was 1,882 yards.

In which year did the NFL leading passer throw 24 touchdowns? Find the "touchdown" column. Read down this column until you find 24. Look across the row to the "year" column. Find 1970. The touchdown record was 24 in 1970.

In which year were the most passing attempts made? Look down the "attempts" column. Compare the numbers. The largest is 422. Look left to the "year" column. The answer is 1969.

Try It (Answers on page 268.)

Read each table. Answer the questions. Find the correct answer.

Food Nutritive Value Per Cup

FOOD	WATER %	CALORIES	FAT GRAMS	PROTEIN GRAMS
milk	87	160	9	9
oysters	85	160	4	20
cashews	5	785	64	24
brown sugar	2	820	0	0
grapes	82	65	1	1
lard	0	1850	205	0

1. Which food has the most calories?
   (1) cashews   (2) sugar   (3) lard   (4) milk

2. Which food has no fat?
   (1) brown sugar   (2) grapes   (3) oysters   (4) lard

3. What percent water does lard contain?
   (1) 1850%   (2) 205%   (3) 0%   (4) 100%

4. Which three foods have balanced amounts of fats and proteins?
   (1) milk, oysters, cashews   (2) lard, grapes, brown sugar
   (3) milk, lard, grapes   (4) milk, brown sugar, grapes

5. What percent is grape protein compared to oyster protein?
   (1) 5%   (2) 20%   (3) 2,000%   (4) 50%

Mileage Chart—Five Towns

MILES TO

	Clear Lake	Wheaton	Carver	Blue Spring	Ida Falls
Clear Lake		4	9	13	21
Wheaton	4		7	11	28
Carver	9	7		17	2
Blue Spring	13	11	17		3
Ida Falls	21	28	2	3	

FROM

6. What type of chart is this?

   (1) height   (2) population   (3) area   (4) mileage

7. What do the numbers represent?

   (1) feet   (2) pounds   (3) miles   (4) meters

8. Which two cities are farthest apart?

   (1) Ida Falls, Blue Springs   (2) Wheaton, Ida Falls
   (3) Blue Spring, Carver   (4) Clear Lakes, Ida Falls

9. Which two cities are closest to each other?

   (1) Clear Lake, Wheaton   (2) Blue Spring, Ida Falls
   (3) Carver, Clear Lake   (4) Carver, Ida Falls

10. How far is it from Blue Spring to Carver?

    (1) 21   (2) 13   (3) 11   (4) 17

36. 100% GRAPHS

READING A CIRCLE GRAPH

A circle graph is a circle divided into parts. They are sometimes called pie graphs. Circle graphs show parts in relation to each other and a total sum. They make it easy to show differences in amounts. The entire circle represents the total or 100%. It may also represent $1 or 100 cents. On this type of graph, 25% would appear as 25¢. Look at the example below:

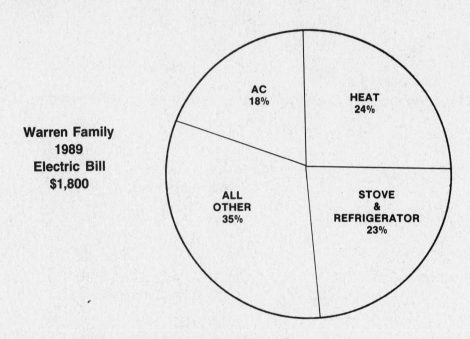

**Warren Family
1989
Electric Bill
$1,800**

Can you interpret this circle graph?

How much did the Warrens spend on air conditioning in 1989? The graph shows that air conditioning (shown as AC) is 18% of the total electric bill. The total bill was $1,800.

$$\text{air conditioning} = 18\% \text{ of } \$1,800$$

$$
\begin{array}{r}
\$1,800 \\
\times \quad .18 \\
\hline
144\,00 \\
180\,0 \\
\hline
\$324.00
\end{array}
$$

The Warrens spent $324 in 1989 on air conditioning.

What did the Warrens spend the most on? How much did they spend on it? The graph shows that the section "all other" is the largest. This section is marked 35%.

"all other" = 35% of $1,800

$$\begin{array}{r} \$1{,}800 \\ \times \quad .35 \\ \hline 9000 \\ 5400 \\ \hline \$630.00 \end{array}$$

The Warrens spent $630 on "all other" electric items in 1989.

Try It (Answers on page 268.)

Read each graph. Answer the questions. Find the correct answer.

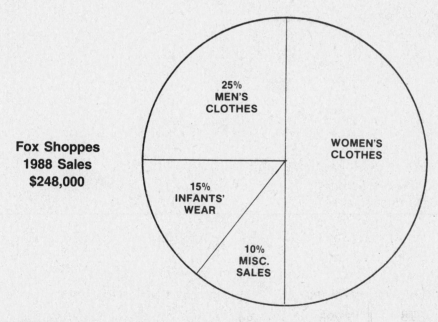

Fox Shoppes
1988 Sales
$248,000

25%
MEN'S
CLOTHES

WOMEN'S
CLOTHES

15%
INFANTS'
WEAR

10%
MISC.
SALES

1. What percent of the circle does women's clothes represent?
   (1) 25%   (2) 50%   (3) 180%   (4) 75%

2. What is the actual dollar amount of sales in women's clothes?
   (1) $128,000   (2) $50,000   (3) $124,000   (4) $125,000

3. What is the actual dollar amount of sales in men's clothing?
   (1) $6,200,000   (2) $62,000   (3) $25,000   (4) $9,920

4. How much infants' wear was sold in 1988?
   (1) $15,000   (2) $37,200   (3) $37,400   (4) $37,000

5. What percent of sales is miscellaneous?
   (1) 50%   (2) 25%   (3) 15%   (4) 10%

Megabuck 1989 Corporate Budget

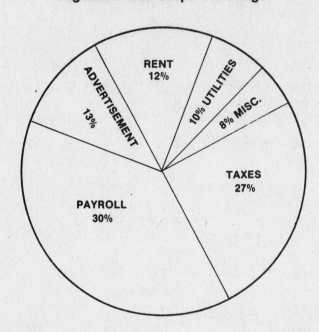

6. What is Megabuck's smallest budget item?
   (1) advertisement   (2) rent   (3) utilities
   (4) miscellaneous

7. What fraction of the budget is utilities?
   (1) $\frac{1}{4}$ (2) $\frac{1}{8}$ (3) $\frac{1}{10}$ (4) $\frac{1}{12}$

8. Out of every dollar, how much does Megabuck spend on miscellaneous?
   (1) 13¢   (2) 8¢   (3) 12¢   (4) 10¢

9. In what area is most of Megabuck Corporation's money spent?
   (1) taxes   (2) advertisement   (3) payroll
   (4) rent

10. Megabuck's total budget for 1989 was $8 million. How much of this was spent for taxes?
    (1) $2,700,000   (2) $216,000   (3) $21,600
    (4) $2,160,000

**The Daily Clarion
Sources of Income
(per dollar)**

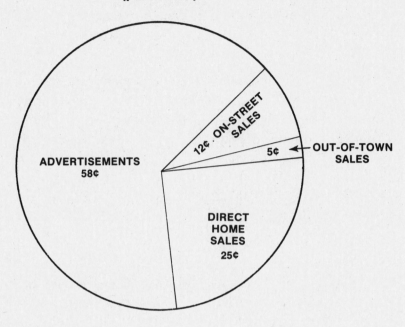

11. Where does this paper make the most money?
    (1) out-of-town sales (2) direct home sales (3) on-street sales
    (4) advertisements

12. What percent of income comes from advertisements?
    (1) 58%   (2) 12%   (3) 5%   (4) 25%

13. What fraction of the income comes from out-of-town sales?
    (1) $\frac{1}{5}$ (2) $\frac{1}{4}$ (3) $\frac{1}{20}$ (4) $\frac{4}{5}$

14. What percent of income is direct home sales?
    (1) 4%   (2) 25%   (3) $\frac{1}{4}$%   (4) 40%

15. The Daily Clarion's total annual income was $160,000 last year.
    How much of their income came from out-of-town sales?
    (1) $80   (2) $8,000   (3) $800   (4) $80,000

READING A DIVIDED BAR GRAPH

Another form of a 100% graph is a divided bar graph. One entire bar
represents the whole amount, 100%. The bar is divided into parts.
Each part shows a section of the whole.

AZ Corp. Revenues
$240,000

REPAIR SERVICES 30%	PARTS SALES 20%	NEW SALES 50%

How much of AZ Corporation's revenues were from new sales? You can see that the new sales part of the graph is 50%. 50% is $\frac{1}{2}$ the length of the bar. This means that $\frac{1}{2}$ of the revenues are due to new sales.

$$\begin{array}{r} 240,000 \\ \times \quad\quad .50 \\ \hline 120,000.00 \end{array} \quad \text{or} \quad \frac{\overset{120,000}{\cancel{240,000}}}{1} \times \frac{1}{\underset{1}{\cancel{2}}} = \$120,000$$

What percent of AZ Corp.'s revenues came from repair services? 30%

What percent of the revenues came from parts sales? 20%

50% of the revenues from AZ Corporation's new sales came from selling video games. How much money did AZ Corp. make selling video games?

$$50\% \times \frac{1}{2} = 25\%$$

$$\begin{array}{r} 240,000 \\ \times \quad\quad .25 \\ \hline 60,000.00 \end{array} \quad \text{or} \quad 240,000 \times \frac{1}{4} = \$60,000$$

Try It (Answers on page 269.)

Read each graph. Answer the questions. Find the correct answer.

Highlands Senior High School
12th Grade Ethnic Distribution
460 STUDENTS

BLACK 25%	WHITE 50%	HISPANIC 25%

1. What percent of the 12th grade is Black?
   (1) 50%   (2) 2.5%   (3) 20%   (4) 25%

2. Which two ethnic groups are equally represented in the 12th grade?
   (1) White, Hispanic   (2) Black, White   (3) Black, Hispanic
   (4) cannot tell

3. What percent of the 12th grade is White?
   (1) 50%   (2) 25%   (3) 5%   (4) 75%

4. How many White students are there in the 12th grade?
   (1) 460   (2) 115   (3) 230   (4) 345

5. How many Hispanic students are there in the 12th grade?
   (1) 115   (2) 460   (3) 320   (4) 345

**Payroll Deduction
Distribution**

```
┌─────────────────┐
│   WITH-         │
│   HOLDING       │
│   TAX           │
│   67.5%         │
│                 │
│                 │
├─────────────────┤
│  F.I.C.A. 12%   │
├─────────────────┤
│  HOSP. 9.5%     │
├─────────────────┤
│  PAYROLL        │
│  SAVINGS 6%     │
├─────────────────┤
│  UNION 5%       │
└─────────────────┘
```

6. Out of each dollar withheld, how much is for withholding taxes?
   (1) 12¢   (2) $9\frac{1}{2}$¢   (3) 6¢   (4) $67\frac{1}{2}$¢

7. F.I.C.A. (Federal Insurance Contribution Act), known as Social Security, is twice as much as what other deduction?
   (1) payroll savings   (2) withholding   (3) union
   (4) hosp.

8. If all the deductions amounted to $36, how much would go for withholding tax?
   (1) $22.50   (2) $4.32   (3) $24.30   (4) $31.50

9. Which area represents the smallest deduction?
   (1) F.I.C.A.   (2) hosp.   (3) union   (4) payroll savings

10. One week the total deductions for an employee amounted to $36. How much of this deduction was for hospitalization?
    (1) $9.50   (2) 95¢   (3) 34.2¢   (4) $3.42

37. PICTOGRAPHS

READING A PICTOGRAPH

A pictograph uses symbols or pictures to show amounts. Each symbol or picture usually equals a certain amount. The symbol used in the pictograph should always be clearly defined. In the case of large numbers, the amount the symbols represents is often rounded off. Pictographs are not as accurate as other graphs. They are used for strong visual impact. Look at this example:

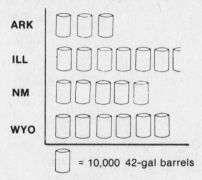

Crude Petroleum Production 1987

= 10,000 42-gal barrels

Can you read this pictograph? How much does each barrel stand for? Below the graph is the key:

= 10,000 42-gallon barrels.

Each barrel on the pictograph stands for 420,000 gallons of oil.
Compare Arkansas's production to Wyoming's production:

$$\text{Write the ratio: } \frac{ARK}{WYO} = \frac{3}{6} = \frac{1}{2} = 50\%$$

Arkansas produces 50% as much crude as Wyoming.

How many barrels of crude did Illinois produce in 1987? There are $6\frac{1}{2}$ barrels shown in Illinois. Illinois produced $6.5 \times 10,000 = 65,000$ barrels of crude in 1987.
How many gallons did Illinois produce?

$$1 \text{ barrel} = 42 \text{ gallons}$$

$$65,000 \times 42 = 2,730,000 \text{ gallons.}$$

251

Try It (Answers on page 269.)

Read each graph. Answer the questions. Find the correct answer.

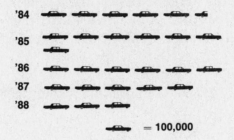

Zeta Corp. Auto Exports

1. What is the title of the graph?
   (1) Auto Exports
   (2) Zeta Auto Exports
   (3) Zeta Corp.
   (4) Zeta Corp. Auto Exports

2. What does the vertical scale represent?
   (1) sales   (2) 100,000   (3) cars   (4) years

3. What amount does each symbol stand for?
   (1) 100,000   (2) 10,000   (3) 1,000   (4) 1

4. How many autos did Zeta export in 1985?
   (1) 70,000   (2) 7,000   (3) 7   (4) 700,000

5. How many autos did Zeta export in 1984?
   (1) 5,000   (2) 55,000   (3) 550,000   (4) 500

Tri-County Hog Production 1986

= 1,000

6. What amount does each symbol stand for?
   (1) 1   (2) 10   (3) 100   (4) 1,000

7. What counties are represented on the graph?
   (1) Lee   (2) Davis   (3) Cook
   (4) Lee, Davis, Cook

8. Which county produced the most hogs in 1986?
   (1) Lee   (2) Cook   (3) Davis
   (4) cannot determine

9. Rank the counties in hog production, largest first.
   (1) Cook, Davis, Lee (2) Davis, Lee, Cook (3) Lee, Cook, Davis
   (4) Cook, Lee, Davis

10. How many hogs were produced in Cook County?
    (1) $3\frac{1}{2}$ (2) 35   (3) 3,500   (4) 350

38. BAR GRAPHS

READING A BAR GRAPH

Bar graphs use bars to show data. The lengths of the bars clearly show differences in data. Comparisons of many items can be made quickly.

There are two types of bar graphs, horizontal and vertical. The bars on a horizontal graph run sideways. The bars on a vertical graph run up and down. The data may be the same. If each bar represents the same information for different years, you can read the graph for trends. If the bars grow in length, they show an upward trend. If they become shorter, they show a downward trend.

California
Non-Motor Vehicle Accidental Deaths
1989

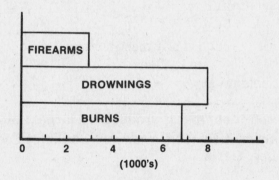

This is a horizontal bar graph. The title tells us what the graph shows: the number of non-motor vehicle accidental deaths in California for 1989. There were three causes—firearms, drownings, and burns. The lengths of the bars show what caused the most number of accidental deaths in California in 1989. Can you interpret this graph?

How many deaths were due to drownings? First, find the bar showing drownings. This bar ends at the number 8. The horizontal scale at the bottom of the graph is in 1,000's. This means that 8 stands for 8,000. There were 8,000 deaths due to drownings in California in 1989.

254

READING BETWEEN THE LINES

Below is a vertical bar graph for the same data. The horizontal scale now runs vertically.

How many deaths were due to firearms? First, find the bar representing firearms. This bar ends halfway between the 2 and the 4. The halfway mark is 3, or 3,000. There were about 3,000 deaths due to firearms in California in 1989.

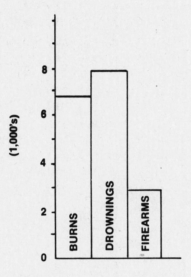

Try It (Answers on page 269.)

Read each graph. Answer the questions. Find the correct answer.

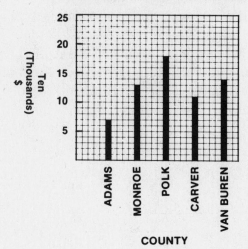

1990 State Funds For Roads

1. What do the numbers on the vertical scale represent?
   (1) $1,000   (2) $10,000   (3) $100,000   (4) $1,000,000

2. Which county received the most in state funds for roads?
   (1) Polk   (2) Monroe   (3) Van Buren   (4) Carver

3. How much did Van Buren County receive in 1990?
   (1) $150,000   (2) $140,000   (3) $155,000   (4) $130,000

4. How much more did Van Buren receive than Adams?
   (1) $70,000   (2) $7,000   (3) $700.00   (4) $75,000

5. Which two counties received about the same amount of state funds?
   (1) Monroe-Polk   (2) Adams-Carver   (3) Monroe-Van Buren
   (4) Carver-Van Buren

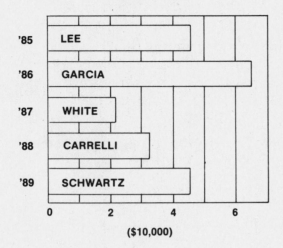

**Sales Leaders for
Widget Corp.**

6. What company do these men work for?
   (1) Sales Co.　　　　　(2) Leaders Corp.　　　(3) Widget Sales
   (4) Widget Corp.

7. What unit is the horizontal scale?
   (1) $10,000　　　(2) years　　　(3) 2　　　　　(4) leaders

8. What does the vertical scale represent?
   (1) $1,000　　　(2) $10,000　　　(3) months　　　(4) years

9. What year experienced the worst decline?
   (1) '85　　　　　(2) '86　　　　　(3) '87　　　　　(4) '88

10. What is the trend from 1987 to 1989?
    (1) decreasing　　　　(2) neutral　　　　　　(3) increasing
    (4) cannot determine

11. What were the sales for the leader in 1986?
    (1) $65,000　　　(2) $75,000　　　(3) $72,500　　　(4) $70,000

12. Garcia's sales are approximately what percent of Carrelli's?
    (1) 50%　　(2) 400%　　(3) 200%　　(4) 150%

13. What is the maximum amount that can be shown on the graph?
    (1) $60,000　　　(2) $70,000　　　(3) $65,000　　　(4) no maximum

14. Which two years had the same amount for the sales leader?
    (1) '85, '86   (2) '85, '87   (3) '85, '88   (4) '85, '89

15. Exactly what type of graph is this?
    (1) line graph   (2) bar graph
    (3) vertical bar graph   (4) horizontal bar graph

Adult High School Diplomas Issued City High

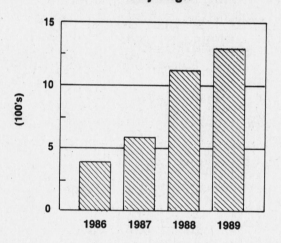

16. What type of graph is this?
    (1) horizontal bar   (2) vertical line   (3) vertical bar
    (4) horizontal line

17. What do the numbers on the horizontal scale represent?
    (1) 5's   (2) 100's   (3) 500's   (4) years

18. Approximately how many adult diplomas did City High issue in 1989?
    (1) 1,000   (2) 1,200   (3) 1,250   (4) 1,300

19. In which years were over 1,000 adult high school diplomas issued at City High?
    (1) '88, '89   (2) '87, '88   (3) '87, '89   (4) '86, '87

20. What can you say about the future for adult study at this school?
    (1) decreasing   (2) about the same   (3) increasing
    (4) neutral

39. LINE GRAPHS

A line graph shows changing information in a visual form. A line graph can show how temperature goes up and down or how the price of an item rises and falls. A line going up shows an increase. A line going down shows a decrease. If a line rises steadily, it shows an upward trend. One that falls steadily shows a downward trend.

A line graph can show a comparison of changing information for more than one thing. For example, two lines on one graph can represent the total sales figures for two different companies.

READING A LINE GRAPH

Every line graph should have a title and two scales. The title tells what is being measured. The scales tell how the information is being measured. A horizontal scale goes across the graph. It often measures time in hours, days, months, or years. A vertical scale goes up and down the side of a graph. It often measures temperature or money. Before you read any line graph, be sure that you understand the title and the two scales of the graph. Look at this example:

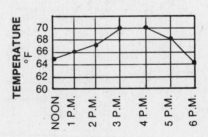

The title tells us what the graph shows: the temperature from noon to 6 P.M. on Monday.

The horizontal scale of this graph shows hours from noon to 6 P.M. The vertical scale shows temperature in °F (degrees in Fahrenheit).

What was the temperature at 3 P.M.? Find 3 P.M. on the horizontal scale. Read up to the dot. From the dot, read left along the line to the vertical scale. The vertical scale number is 70. It was 70° Fahrenheit at 3 P.M.

At what time was it 66°? This is the opposite of the first question. First, find 66° on the vertical scale. Read over to the dot. Read down from the dot to the horizontal scale. The horizontal scale number is 1 P.M. It was 66° at 1 P.M.

Reading Between the Lines

Sometimes a dot falls between two lines on a graph. You must estimate the answer.

What was the temperature at 2 P.M.? Read up from 2 P.M. to the dot. The dot does not fall exactly on a line. It is between the lines for 66° and 68°. It is about halfway between the lines. You can estimate the temperature to be 67°.

Try It (Answers on page 270.)

Read each graph. Answer the questions. Find the correct answer.

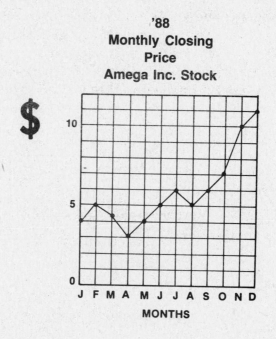

'88
**Monthly Closing
Price
Amega Inc. Stock**

1. What does this graph show?
   (1) months
   (2) dollars
   (3) dates
   (4) monthly closing price

2. What do the letters at the bottom of the graph represent?
   (1) days   (2) weeks   (3) months   (4) closings

3. What do the numbers on the vertical scale represent?
   (1) years   (2) months   (3) dollars   (4) dates

4. What was the price of Amega stock in March?
   (1) $4   (2) $4.50   (3) $5   (4) $5.50

5. How many monthly decreases are shown on the graph?
   (1) 8   (2) 2   (3) 3   (4) 4

6. In what month was the biggest increase?
   (1) October   (2) January   (3) April   (4) August

7. What is the general trend from mid-year on?
   (1) increasing   (2) decreasing   (3) stable   (4) cannot determine

8. What was the lowest selling price for 1988?
   (1) $3   (2) $11   (3) $4   (4) $3.50

9. What was the highest selling price for 1988?
   (1) $10   (2) $10.50   (3) $12   (4) $11

10. What percent of the November selling price was the February
    selling price?
    (1) 200%   (2) 10%   (3) 50%   (4) 25%

**American League
Home Run Leaders
1960-1970**

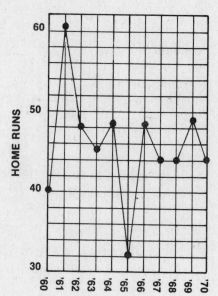

11. What do the numbers on the vertical scale represent?
    (1) dollars   (2) years
    (3) number of leaders   (4) number of home runs

12. What do the numbers on the horizontal scale represent?
    (1) years   (2) home runs   (3) leaders
    (4) cannot determine

13. In what year were the most home runs hit?
    (1) 1960   (2) 1961   (3) 1963   (4) 1965

14. In what year were the least home runs hit?
    (1) 1961   (2) 1960   (3) 1965   (4) 1964

15. How many home runs did the leader hit in 1961?
    (1) 60   (2) 40   (3) 61   (4) 48

16. How many home runs did the leader hit in 1963?
    (1) 50   (2) 49   (3) 48   (4) 45

17. What year showed the greatest decline?
    (1) '60   (2) '61   (3) '64   (4) '65

18. What was the percent increase in home runs hit by the leaders
    between 1960 and 1961? Round off to the nearest percent.
    (1) 67%   (2) 64%   (3) 68%   (4) 53%

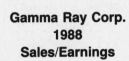

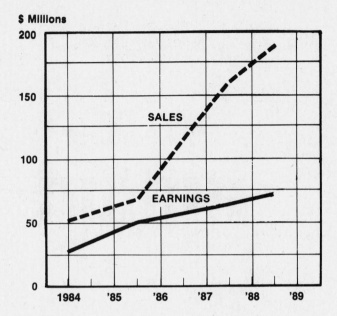

19. What were Gamma Ray's sales in 1988?
    (1) $74 million   (2) $175 million   (3) $101 million
    (4) $200 million

20. What were Gamma Ray's earnings in 1988?
    (1) $74 million   (2) $70 million   (3) $175 million
    (4) $160 million

21. How much greater were sales than earnings in 1988?
    (1) $101 million   (2) $104 million   (3) $106 million
    (4) $105 million

22. In what year was the difference in sales and earnings the smallest?
    (1) '84   (2) '85   (3) '89   (4) '88

23. What was the difference in sales and earnings in 1984?
    (1) $25 million   (2) $20 million   (3) $30 million
    (4) $15 million

24. In 1984 about what percent of sales were earnings?
    (1) 200%   (2) 50%   (3) 40%   (4) 25%

25. In what year did sales first exceed $100 million?
    (1) '84   (2) '85   (3) '86   (4) '87

40. MEASURING YOUR PROGRESS

Read each graph. Answer the questions. Find the correct answer.

**Technical University
Engineering Graduates**

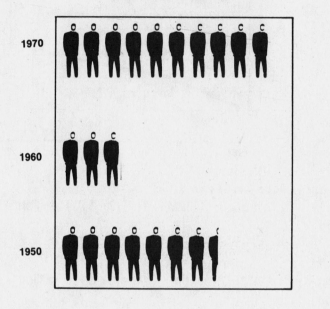

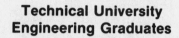

1. How many graduates does each symbol represent?
   (1) 100   (2) 10   (3) 1,000   (4) 1

2. How many students graduated in 1970?
   (1) 100   (2) 10,000   (3) 1,000   (4) 10

3. The number of 1960 graduates is what percent of the number of 1970 graduates.
   (1) $\frac{3}{10}$%   (2) 3%   (3) 70%   (4) 30%

**Comparative
Stock Prices
A Corp./Z Corp.**

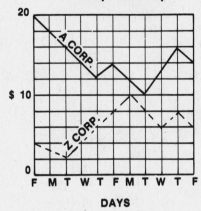

4. On which day did A Corp. first double Z Corp.?
   (1) Thursday   (2) Tuesday   (3) Monday
   (4) Wednesday

5. What was A Corp.'s stock price on Thursday of the second full
   week shown?
   (1) $13   (2) $16   (3) $8   (4) $4

6. On which day were the stock prices of the two corporations
   farthest apart?
   (1) Monday   (2) Tuesday   (3) Wednesday
   (4) Friday

Charge Account Payment Schedule

If The New Balance is:	The Minimum Payment will be:
$ .01 to $ 10.00	Balance
10.01 to 200.00	$10.00
200.01 to 250.00	15.00
250.01 to 300.00	20.00
300.01 to 350.00	25.00
350.01 to 400.00	30.00
400.01 to 450.00	35.00
450.01 to 500.00	40.00
Over $500.00	1/10 of New Balance

7. What percent of the balance is your payment if your charge account balance is over $500?
   (1) 1%   (2) 10%   (3) $\frac{1}{10}$ %   (4) $50

8. How much will your payment be if your charge account balance is $5.95?
   (1) 0   (2) 59.5¢   (3) 60¢   (4) $5.95

9. How much is the payment on a balance of $329.49?
   (1) $20   (2) $25   (3) $30   (4) $32.95

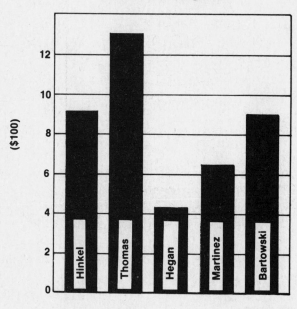

**Subscription Sales
Worldwide Magazine**

10. Which salesperson was next to last in sales?
    (1) Hinkel   (2) Martinez   (3) Hegan
    (4) Bartowski

11. How much did the leader sell?
    (1) $13,000   (2) $130   (3) $1,400   (4) $1,300

12. Which two salespersons tied in sales?
    (1) Hegan-Hinkel   (2) Hinkel-Martinez   (3) Hinkel-Thomas
    (4) Bartowski-Hinkel

1989
Miller Hardware Sales
$320,000

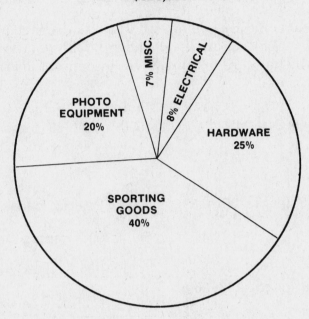

13. How much were hardware sales in 1989?
    (1) $25,000   (2) $8,000   (3) $80,000   (4) $800

14. Compare the sales of electrical to sporting goods.
    (1) 20%   (2) 500%   (3) 200%   (4) 40%

15. What is the difference in sales between electrical and miscellaneous?
    (1) $22,400   (2) $25,600   (3) $32,000   (4) $3,200

Check your answers on page 270.

ANSWERS

HOW MUCH DO YOU KNOW ALREADY?, PAGE 239

Put an X by each problem you missed. Put an X by each problem you did not do. The problems are listed under the titles of the chapters in this book. If you have <u>one or more Xs</u> in a section, study that chapter to improve your skills.

CHARTS (pages 241–243)

1. **43**

100% GRAPHS (pages 244–250)

2. **25%**

3. **30**

PICTOGRAPHS (pages 251–253)

4. **1,800**

BAR GRAPHS (pages 254–258)

5. **4,200 ft. approximately**

LINE GRAPHS (pages 259–264)

6. **Nov.–Dec.**

TRY IT, PAGE 241

1. **(3) lard**

2. **(1) brown sugar**   3. **(3) 0%**

4. **(4) milk, brown sugar, grapes**   5. **(1) 5%** $\frac{1}{20} = .05 = 5\%$

6. **(4) mileage**

7. **(3) miles**

8. **(2) Wheaton and Ida Falls**

9. **(4) Carver and Ida Falls**

10. **(4) 17 miles**

TRY IT, PAGE 245

1. **(2) 50% $\frac{1}{2}$ circle**

2. **(3)** 50% of $248,000 = **124,000**

3. **(2)** $\begin{array}{r} \$248,000 \\ \times \quad\quad .25 \\ \hline \$62,000.00 \end{array}$

4. **(2)** $\begin{array}{r} 248,000 \\ \times \quad\quad .15 \\ \hline \$37,200 \end{array}$

5. **(4) 10%**

6. **(4) miscellaneous**

7. **(3)** $10\% = \frac{10}{100} = \frac{1}{10}$ 8. **(2)** **8¢** $= 8\%$ of $1.00

9. **(3) payroll**—30%   10. **(4)**
$$\begin{array}{r} 8,000,000 \\ \times .27 \\ \hline \$2,160,000 \end{array}$$

11. **(4) advertisements**   12. **(1)** 58¢/dollar $= \frac{58}{100} =$ **58%**

13. **(3)** $\frac{5}{100} = \frac{1}{20}$ 14. **(2)** 25¢/dollar = **25%**

15. **(2)** $8,000 $\frac{1}{\cancel{20}} \times \frac{\cancel{160,000}}{1} =$ **$8,000**

TRY IT, PAGE 248

1. **(4) 25%**   2. **(3) Black, Hispanic**   3. **(1) 50%**   4. **(3)**
$$\begin{array}{r} 4\ 60 \\ \times .50 \\ \hline 2\ 30.00 \end{array}$$

5. **(1)**
$$\begin{array}{r} 460 \\ \times\ .25 \\ \hline 115 \end{array}$$
6. **(4)** $67\frac{1}{2}$**¢**   7. **(1) payroll savings**

8. **(1)** $36 $\times$ .675 = **$24.30**   9. **(3) union**   10. **(4)** $36 $\times$ .095 = **$3.42**

TRY IT, PAGE 252

1. **(4) Zeta Corp. Auto Exports**   2. **(4) years**   3. **(1) 100,000 cars**

4. **(4) 700,000**   5. **(3) 550,000**   6. **(4) 1,000 hogs**

7. **(4) Lee, Davis, Cook**   8. **(1) Lee**   9. **(3) Lee, Cook, Davis**

10. **(3) 3,500**

TRY IT, PAGE 255

1. **(2) $10,000**   2. **(1) Polk**   3. **(2) $140,000**

4. **(1) $70,000**   5. **(3) Monroe-Van Buren**   6. **(4) Widget Corp.**

7. **(1) $10,000**   8. **(4) years**   9. **(3) 1987**

10. **(3) increasing**   11. **(1)** Garcia—**$65,000**   12. **(3)** $\frac{65,000}{32,500} = \frac{2}{1} = \textbf{200\%}$

13. **(2) $70,000**   14. **(4) 1985, 1989**   15. **(4) horizontal bar graph**

16. **(3) vertical bar**   17. **(4) years**   18. **(3)** approximately **1,250**

19. **(1) 1988, 1989**   20. **(3) increasing**

TRY IT, PAGE 260

1. **(4) monthly closing price**   2. **(3) months**   3. **(3) dollars**

4. **(2) $4.50**   5. **(3) 3,** February, March July   6. **(1) October**

7. **(1) increasing**   8. **(1) $3,** April   9. **(4) $11,** December

10. **(3)** $\frac{5}{10} = \frac{1}{2} = \textbf{50\%}$ 11. **(4) number of home runs**   12. **(1) years**

13. **(2) 1961**   14. **(3) 1965**   15. **(3) 61**

16. **(3) 48**   17. **(4) 1965**

18. **(4)** $\begin{array}{r} 61 \\ -\ 40 \\ \hline 21 \end{array}$ $\frac{21}{40} = 52.5\% = \textbf{53\%}$

19. **(2) $175 million**   20.   **(2)** approximately **$70 million**

21. **(4)** approximately **$105 million**   22. **(2) '85**

23. **(1)** approximately **$25 million**

24. **(2)** $\dfrac{\text{earnings}}{\text{sales}} = \dfrac{\$25\ \text{million}}{\$50\ \text{million}} = \frac{1}{2} = \textbf{50\%}$

25. **(3) 1986**

MEASURING YOUR PROGRESS, PAGE 264

1. **(1) 100**   2. **(3)** 1,000   3. **(4)** $\frac{3}{10} = \textbf{30\%}$

4. **(1)** first **Thursday,** A Corp., $12, Z Corp. $6   5. **(2) $16**

6. **(4)** first **Friday,** $16 difference

7. **(2)** 10\% = $\frac{1}{10}$ = **10\%**   8. **(4) $5.95**—balance   9. **(2) $25**

10. **(2) Martinez**   11. **(4) $1,300**   12. **(4) Hinkel-Bartowski,**
                                                      $900 each

13. **(3)** $320,000
              $\times$ .25   or $\frac{\overcancel{80,000}}{\overcancel{320,000}} \times \frac{1}{\underset{1}{\overcancel{4}}} =$ **$80,000**
            ─────────────
              $ 80,000

14. **(1)** $\frac{8}{40} =$ **20%**   15. **(4)**   8% $ 320,0\,00
                                                      $-$ 7% $\times$ .01
                                                    ─────   ─────────────
                                                        1   **$3,200 .00**

POSTTEST

This test is made up of 40 problems similar to the ones in this book. Solve all the problems you are able to. Then check your answers against those on page 276. The results will help you decide whether you should review or continue on in your GED preparation.

1. Write four thousand, four hundred forty-four *rounded to the nearest hundred.*

2. 1,142
   2,011
   +3,411

3. 8,386
   8,279
   3,037
   + 236

4. 2,976
   − 434

5. 42,010
   −6,074

6. 3,986
   × 8

7. 3,571
   × 342

8. $7\overline{)4452}$

9. $622\overline{)60,600}$

10. Reduce to lowest terms: $\frac{6}{32}$

11. $\frac{2}{7}$
    $3\frac{3}{7}$
    $+1\frac{3}{7}$

12. 11
    $-2\frac{3}{8}$

13. $3\frac{1}{2} - 1\frac{5}{6} =$

14. $2\frac{2}{5} \times 2\frac{1}{2} =$

15. $3\frac{1}{4} \div 2\frac{1}{3}$

16. Which is larger, 9.3 or 9.298?

17. $ .11
    2.99
    + 1.06

18. 12.12
    −1.24

19. $3.60 × 37.5 =

20. $.15\overline{)\$3.45}$

21. Write $\frac{3}{4}$% as a decimal.

22. 102% of 200 =

23. Find the percent of increase from 50 to 65.

24. 15% of _____ = 6

25. How much must be repaid on a 3-month loan of $2,500 at 12% annual interest?

26. According to the chart below, how much does it cost to send a 2.4 lb. package to Zone 3?

Parcel Post Rate Schedule

Weight		Zones		
from	**to**	**Local**	**1 & 2**	**3**
1 lb.	2 lbs.	$1.15	$1.35	$1.39
2 lbs.	3 lbs.	1.23	1.45	1.53
3 lbs.	4 lbs.	1.29	1.50	1.65
4 lbs.	5 lbs.	1.36	1.66	1.77

27. According to the circle graph (*right*), what percent of U.S. freight is transported by water routes?

HOW U.S. FREIGHT IS MOVED

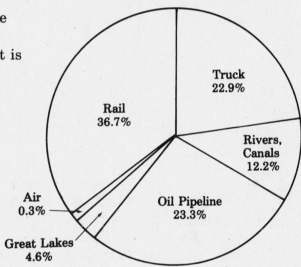

28. According to the pictograph below, do Arkansas and Indiana, together, produce more or fewer eggs than California?

EGG PRODUCTION IN A YEAR

Arkansas ⬭⬭⬭⬭

California ⬭⬭⬭⬭⬭⬭⬭⬭

Georgia ⬭⬭⬭⬭⬭

Indiana ⬭⬭⬭⬭

⬭ = 1 Billion Eggs

29. According to the bar graph
(*right*), what was the first year
the U.S. population was more
than double its size in 1900?

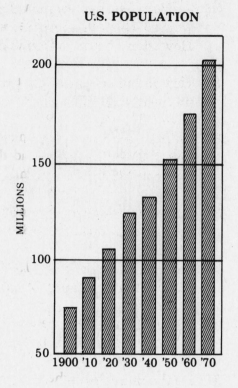

U.S. POPULATION

30. In what year, according to the
line graph (*right*), was the
Safety & Health Administra-
tion's budget twice as large as
it had been the year before?

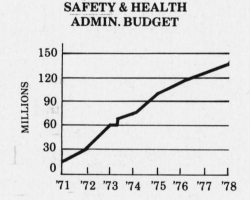

**SAFETY & HEALTH
ADMIN. BUDGET**

31. An airplane hit an air pocket and fell from 20,100 feet altitude to
19,800 feet. How far did it fall?

32. Verva can purchase new living room furniture by paying $1500
outright or by paying $75 a month for two years. How much will
the furniture cost if she buys it on time?

33. Jackson worked overtime three days one week. Two days he worked $1\frac{1}{4}$ hours overtime, and the third day he worked $1\frac{3}{4}$ hours. How many hours overtime did he work that week?

34. How many ounces of ground turkey should Luis buy to make 40 meatbealls that weigh $2\frac{1}{4}$ ounces each?

35. Paul jogged three mornings one week. He ran 1.5 miles the first day, 2.25 miles the second day, and 4.3 miles the last day? How many miles did he run that week?

36. A 6.8-pound roasting chicken normally sells for $1.39 per pound. On sale the chicken cost $5.10. What was the per pound sale price?

37. Janet tipped her waiter 15% of the cost of her dinner. The tip was $1.05. How much did Janet's meal cost?

38. After Christmas a treetop decoration was marked down from $12 to $9. By what percent did the price decrease?

39. Do the numbers on the bars in the graphs below represent percents or amounts?

40. According to the graphs below, in which portion of the United States did the population NOT grow from 1972 to 1979?

U.S. POPULATION in millions

	West	North Central states	South	Northeast
1972	36	57	65	50
1976	39	58	69	49
1979	41	58	72	49

POSTTEST ANSWERS

Put an X by each problem you did wrong or did not do. For each problem you missed, review the chapter indicated in the table below. If you got all of the problems correct, you are probably ready to go on to further GED preparation.

Group	Computation Problems	Word Problems
Whole Numbers	.1. **4,400** 2. **6,564** 3. **19,938** 4. **2,542** 5. **35,936** 6. **31,888** 7. **1,221,282** 8. **636** 9. **97 r 266**	31. **300 feet** 32. **$1800**
Fractions	10. $\frac{3}{16}$ 11. $5\frac{1}{7}$ 12. $8\frac{5}{8}$ 13. $1\frac{2}{3}$ 14. **6** 15. $1\frac{11}{28}$	33. $4\frac{1}{4}$ **hours** 34. **90 ounces**
Decimals	16. **9.3** 17. **$4.16** 18. **10.88** 19. **$135.00** 20. **23**	35. **8.05 miles** 36. **$.75/pound**
Percents	21. **.0075** 22. **204** 23. **30%** 24. **40** 25. **$2575**	37. **$7** 38. **25%**
Charts and Graphs	26. **$1.53** 27. **16.8%** 28. **Fewer** 29. **1950** 30. **1973**	39. **Amounts** 40. **Northeast**

You will find instruction about problems like those on the Posttest in the chapters indicated in the following table.

Problem	Chapter	Problem	Chapter	Problem	Chapter
1	2	11, 33	14	21	28
2	3	12	15	22	29
3	4	13	16	23, 38	30
4	5	14, 34	17	24, 37	31
5, 31	6	15	18	25	32
6	7	16	21	26	35
7, 32	8	17, 35	22	27, 39, 40	36
8	9	18	23	28	37
9	10	19	24	29	38
10	13	20, 36	25	30	39